Digital Circuits

Study Guide and Laboratory Manual

William J. Streib

South Holland, Illinois

THE GOODHEART-WILLCOX COMPANY, INC.

Publishers

INTRODUCTION

This workbook and laboratory manual is keyed to the textbook DIGITAL CIR-CUITS by William J. Streib. It can, however, be used effectively with other textbooks.

Directions related to the use of this manual will be supplied by your instructor. He or she has designed your course to meet certain objectives. As a result, your instructor can best fit this manual into your learning experience.

Your instructor will also describe the safety procedures to be followed in the laboratory. If you are unsure about how to handle a safety-related situation, ask your instructor at once.

In this manual, laboratory assignments do not contain detailed instructions. Rather, you are told only what is to be accomplished. This approach was selected to approximate such tasks in industry.

During the first few experiments, this lack of detailed instructions may bother you. You will, however, quickly develop the necessary skills and learn to organize your efforts to complete the task at hand.

Experiments in this manual emphasize three concepts:

1. Workers in the field of digital electronics must possess both technical knowledge and technical skill. The ratio of knowledge to skill depends on individual job assignments. Engineers usually need large amounts of technical knowledge and comparatively little technical skill. Electricians must be highly skilled. However, all workers in the digital electronics field must have both technical knowledge and technical skill to be effective on the job.

Experiments in this manual will aid in developing both your knowledge and skill. The skills emphasized are related to experimental circuit construction, troubleshooting, and digital measurements. Analytical thinking is emphasized, along with developing clear statements of a problem and its possible solutions.

2. Digital theory is a tool to be used in the design and troubleshooting of practical circuits. Experiments have been selected to prove that digital theory does indeed describe the actions of real circuits.

3. If you cannot predict the signal that should be present at a point in a circuit, electrical measurements made at that point will be of little value. It is only by comparing measured and predicted values that information about a circuit's operation is obtained.

In each experiment, you will be asked to predict signals within a circuit. You will then make the necessary measurements. The circuit's action is then tested by comparing the predicted and measured results.

CONTENTS

DEVELOPING EFFECTIVE EXPERIMENTAL CIRCUIT WIRING TECHNIQUES

Methods of quickly and effectively wiring experimental circuits are described in this section. Attention is paid to safety and the saving of work. Learning takes place step by step.

BREADBOARDING

The term *BREADBOARDING* implies the building of experimental circuits. In the early days of radio, circuits were often built on wooden boards. It was said that many radio amateurs used their family's cutting boards (called breadboards in those days) as bases for their projects. Today, this colorful term is an accepted part of the vocabulary of electronics.

PRESENT-DAY BREADBOARDS

Special sockets have been developed to aid in breadboarding. These permit the assembly of circuits without the use of solder.

Fig. 1 shows a widely used system. To avoid trade names, breadboards of this sort will be called strip sockets.

USE OF STRIP SOCKETS

Beneath each set of five holes in a strip socket like that in Fig. 1 is a spring contact. Wires inserted in a given five-hole set are electrically connected through these contacts. See Fig. 2.

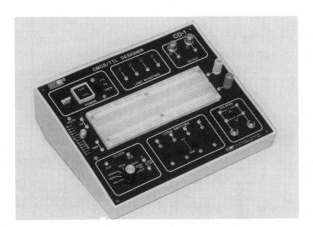

Fig. 1. Strip sockets are widely used in schools and in industry to breadboard circuits. The strip socket on the trainer shown permits the construction of complex digital circuits. (E&L Instruments, an Interplex Electronics Company)

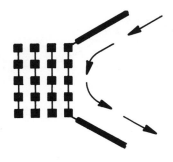

Fig. 2. Wires placed in any five-hole set in a strip socket are electrically connected.

WIRE

Oversize wire can damage strip sockets. Wire should be 0.033 inches in diameter or smaller. That is, wire larger than 20 gage must be avoided. If you do not know if a given wire is number 20 or smaller, use the general rule, a wire that must be forced into a socket is too large.

Number 22, solid, insulated wire is recommended. Stranded wire should not be used. Strands may break off and lodge in the socket.

When parts with large leads must be used, the methods in Fig. 3 are suggested. At the top, small diameter wires have been soldered to the leads. At the right, binding posts found on many trainers have been used.

WIRE PREPARATION

Quality wire cutters should be used. Burrs (rough ends on wires) can damage sockets.

About 1/4 in. of insulation should be removed from the ends of wires. Problems result if too much insulation is removed. In some sockets, internal short circuits are likely if too much wire is inserted. On the outside top area of sockets, exposed wires result in short circuits between adjacent wires.

Wire strippers should be adjusted to avoid nicking wires. Such wires tend to break and leave bits of wire in socket holes.

INSERTING ICs

Strip-socket holes are spaced to match pin spacing on DIPs. Fig. 3 shows a DIP mounted on a strip socket. Connections to a pin are made through the remaining four holes in a five-hole set. That is, up to four wires can be directly connected to any DIP pin.

When a DIP is new, DIP leads flare outward, Fig. 4. As a result, new chips may be difficult to insert. To overcome this problem, place the chip on its side as shown, and bend its pins inward. Insertion tools that correct for flare are available.

Integrated circuits that have been used in student laboratories often have bent pins. Long-nose pliers can be used to straighten these.

Extraction tools should be used to remove DIPs. It is nearly impossible to remove DIPs by hand without bending pins. If a tool is not available, a small screwdriver can be used. The screwdriver should be slid under the DIP package at either of its ends. By prying up alternate ends, DIPs can be removed without damage.

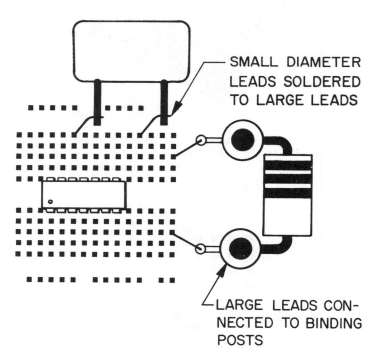

SMALL DIAMETER LEADS SOLDERED TO LARGE LEADS

LARGE LEADS CONNECTED TO BINDING POSTS

Fig. 3. The methods shown may be used to avoid inserting leads larger than number 20 into strip sockets.

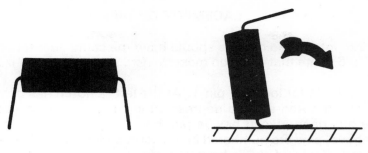

Fig. 4. Leads on new DIPs flare outward and must be straightened to fit easily into strip sockets.

BUSES

One or more buses are provided at the top and bottom of most strip sockets. Return to Fig. 1. In some cases, buses are available as detachable strips, Fig. 5.

Although placed in groups of five, all bus holes are connected. An exception is found in sockets with screw holes in the middle of the bus area.

To connect the two halves of such buses, jumpers (short wires) are used. Snap fittings, interlocking ears, or screws hold buses in place.

WIRING SUGGESTIONS

The following suggestions are aimed at producing circuits that are easy to troubleshoot. Few complex circuits are assembled without one or more errors or faulty components. Often, the time required to find these faults is longer than the assembly time. Because troubleshooting can require a great deal of time, efforts toward building circuits that are easy to troubleshoot are worthwhile. Several topics need to be discussed. These are:
1. Placement of DIPs.
2. Wiring for good appearance and best function.

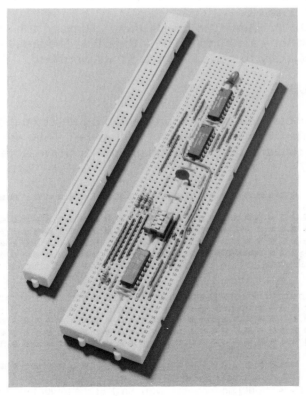

Fig. 5. Some strip sockets are designed so buses can be added. (RSP Electronics Corp.)

PLACEMENT OF DIPs

For easy circuit tracing, integrated circuits should have the same orientation. Pin 1 is usually placed at the lower left. See Fig. 6. Note that chips on most trainers are mounted horizontally (the long direction is horizontal).

Fig. 6 shows three methods for locating pin 1. At the left, a notch has been placed across the end of the DIP. In the middle, a notch-shaped depression in the package serves the same purpose. At the right, a small hole has been placed beside pin 1.

When viewed from above, pins are numbered in the counterclockwise direction. Separate dip sockets 1/4 in. high can be used. Some of these have the pin numbers marked.

WIRING FOR GOOD APPEARANCE AND BEST FUNCTION

The following suggestions result in circuits that have a professional appearance. However, this is not the purpose of these suggestions. They are intended to produce circuits that are easy to troubleshoot.

INSTALL WIRES AROUND, NOT OVER

It is usually better to go around, rather than over, a DIP. Return to Fig. 6. When troubleshooting, DIPs must be removed for testing and replacement, so wires that span chips must be removed. When replaced, such wires are often misconnected. A circuit with two errors is much more difficult to troubleshoot than a circuit with only one error.

LEAD DRESS

The term *LEAD DRESS* describes the placement or routing of wires. Fig. 7 shows three approaches to lead dress on a strip socket.

In the first circuit, leads have been dressed against the socket. Corners have been made square. This layout aids troubleshooting.

The second circuit is almost as easy to troubleshoot as the first. Wires were routed in an orderly way but were not dressed against the socket. Sharp bends were not used at corners, but crossed wires were avoided when possible.

The wiring of the last circuit hinders troubleshooting. Wires of the same length were used on all runs. They often cross, and they are bunched near the DIPs. The small amount of time saved during the wiring of this circuit is likely to be lost during troubleshooting.

POWER AND GROUND BUSES

All IC chips must be connected to a power source and ground. Since these leads do not enter directly into the logic of a circuit, they should be kept out of the way. They are usually connected first, and buses on sockets are used to distribute power and ground to chips. Note the power and ground leads in the first circuit of Fig. 7.

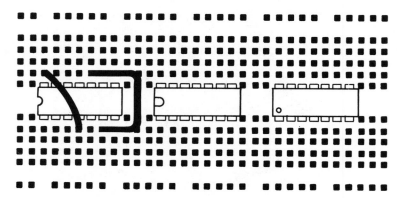

Fig. 6. To aid in troubleshooting, DIPs should be mounted with pin 1 at the lower left. Space should be left between DIPs so leads can go around rather than over DIPs.

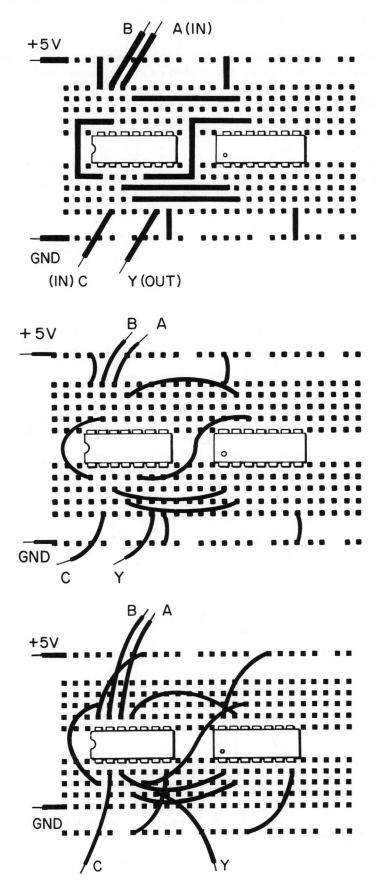

Fig. 7. Electrically, the displayed drawings represent identical circuits. However, the third circuit would be difficult to troubleshoot.

SIGNAL FROM LEFT TO RIGHT

Inputs are usually on the left; outputs are on the right. There are, of course, times when this rule cannot be followed. However, the general direction of signal flow should be to the right.

LOGIC ELEMENTS WITHIN ICs

Fig. 8 is the *PINOUT* (pin outline diagram) of a 7400 integrated circuit. It shows the pins connected to the four logic elements in the device.

When wiring experimental circuits, workers tend to start with the element connected to pins 1 2, and 3. This may not result in the best wiring layout. Because elements within a chip are identical the element that results in the best layout should be used first.

PIN NUMBERS

As a circuit is wired, a record must be kept of the connections that have been made. An effective method of doing this involves recording pin numbers on the logic diagram. See Fig. 9. The resulting diagram is a necessary part of effective troubleshooting.

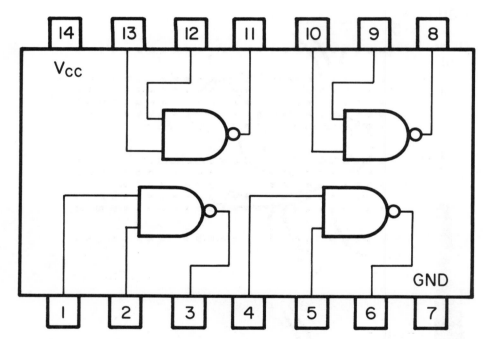

Fig. 8. Pinouts indicate the functions of individual pins.

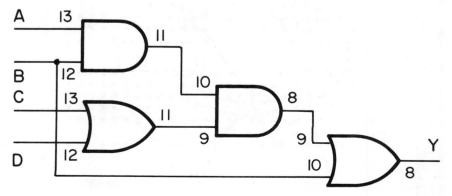

Fig. 9. As a circuit is assembled in the laboratory, pin numbers should be added to its diagram. These numbers are a major aid when troubleshooting.

PLACEMENT OF LEADS

When strip sockets are used, there is a tendency to place leads in holes close to DIPs. This results in crowding and confusion.

Power, ground, and incoming signal leads should be placed in the outermost holes. Long interconnecting leads should use middle holes. Short leads are best placed near DIPs. Again refer to the first circuit in Fig. 7.

JUMPERS

Jumpers (short leads between adjacent holes) may go unnoticed during troubleshooting. They are also difficult to insert and remove. To correct for this, these very short leads should be made slightly longer than necessary.

REUSE OF WIRE

Wires can be reused. They should be carefully removed and stored. Wires with solder on them should not be reused in strip socket holes. The solder end should be cut off and the wire should be restripped.

WIRE AND TEST SUBSECTIONS

Complex circuits are likely to contain errors and faulty parts. Also, the larger a circuit is, the more difficult it is to troubleshoot. To make troubleshooting easier, complex circuits should be built and tested in sections. The time required to test subsections is returned many times over when an error is found before final assembly.

RECORDING AND REPORTING

In present-day industry, technical tasks are seldom complete until they have been documented, Fig. 10. Without a record of what has been done, valuable information is lost.

Fig. 10. A computer is being used to document the results of a repair effort. (Manufacturing Test Div., Hewlett-Packard Co.)

Documenting is a skill. It is best learned through practice. Because reports are best written based on an immediate need (rather than on a request to write a sample report in an English class) report writing is an important part of a laboratory experiment.

Your instructor is your best source of information on the form of documentation to use. He or she is aware of the needs of the industries and educational institutions that recruit from your program. The report form used in the course you are taking should be followed with care.

Chapter 1

DIGITAL TECHNOLOGY

Name _____

Date _____ Score _____

Instructor _____

INTRODUCTION TO WIRING OF EXPERIMENTAL CIRCUITS

1-1. According to this study guide and laboratory manual, why should recommended wiring prac-
tices be used when building experimental circuits?
   a. To improve the appearance of such circuits.
   b. To produce circuits that are easy to troubleshoot.
   c. To speed the assembly of laboratory circuits.

1-2. What do the letters DIP stand for?
   a. Direct-Insertion Package
   b. Dual-In-Line Package
   c. Double-IC Package

1-3. Circle pin 1 on each of these DIPs. Also, draw a square around pin 9 on each package. They
are viewed from above.

a.

b.

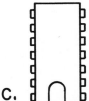

c.

1-4. What name is given to this drawing?
   a. Pinout
   b. Device drawing
   c. Schematic

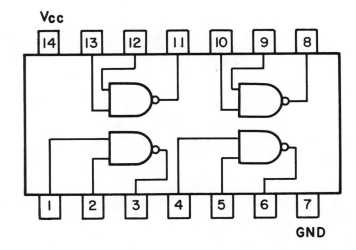

1-5. Based on this drawing of an experimental circuit and the drawing in Problem 4, add pin number to this logic diagram to show how the circuit has been wired.

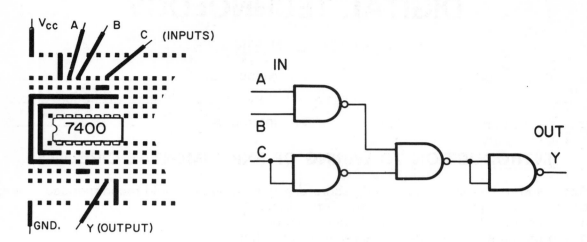

Chapter 1
DIGITAL TECHNOLOGY

Name _____

Date _____ Score _____

Instructor _____

COUNTING IN BINARY

2-1. Complete this table by counting to binary 10,100.

2-2. Indicate the number of bits in each of the following binary numbers.

a. _____   111

b. _____   1,110

c. _____   10,011

2-3. Based on the binary number at the right below, indicate the values (1s and 0s) for the following bits.

a. _____ D2

b. _____ D3   N = 1,011,000

c. _____ D5

DECIMAL	BINARY
0	
1	
2	
3	
4	
5	
6	
7	
8	
9	
10	
11	
12	
13	
14	
15	
16	
17	
18	
19	
20	

2-4. Indicate the place values for the bits of a 9-bit binary number.

D8   D7   D6   D5   D4   D3   D2   D1   D0

__   __   __   __   __   __   __   __   __

2-5. Evaluate (convert to decimal) the following binary numbers.

a.   101 = _____ (decimal)

b. 1,100 = _____ (decimal)

c. 101,011 = _____ (decimal)

2-6. Place a circle around the MSB of this binary number. Then, place a square around its LSB.

100,101,001

Chapter 2
LOGIC ELEMENTS

Name _____

Date _____ Score _____

Instructor _____

SWITCH-BASED AND ELEMENTS

1-1. For each set of switch positions, indicate the logic state of lamp L. If lit, L = 1; if out, L = 0.

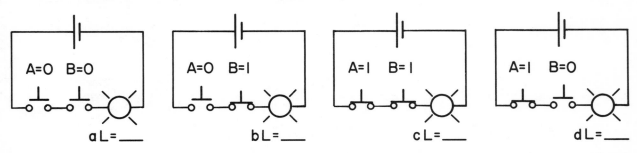

1-2. Complete the symbols for the missing switches in these AND elements for the indicated values of L. If a missing switch can be in either position, show it open.

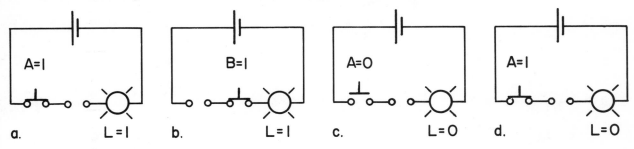

1-3. Complete this truth table for an AND element. The rows are not in standard order.

INPUTS		OUTPUT
A	B	L
1	0	
1	1	
0	1	
0	0	

1-4. Which of the following is the Boolean expression for an AND?

a. A + B = L   b. A/B = L   c. AB = L   d. A − B = L

1-5. Complete each of the following Boolean expressions for an AND.

a. 1 × 0 = _____   b. 0 × 0 = _____   c. 1 × 1 = _____   d. 0 × 1 = _____

1-6. Place a 1 or 0 in each blank to complete the following Boolean expressions. If the missing number may be either a 1 or 0, show a 0.

a. $1 \times \underline{\hspace{1cm}} = 1$ b. $1 \times \underline{\hspace{1cm}} = 0$ c. $0 \times \underline{\hspace{1cm}} = 0$ d. $\underline{\hspace{1cm}} \times 1 = 0$

Chapter 2
LOGIC ELEMENTS

Name _____

Date _____ Score _____

Instructor _____

SWITCH-BASED OR ELEMENTS

2-1. Complete this drawing to show a 2-input, switch-based OR. Use pushbutton switches (A and B) as inputs and lamp (L) as the output.

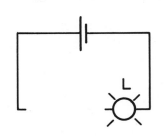

2-2. Complete this truth table for the OR in Problem 1. The rows are not in standard order.

INPUTS		OUTPUT
A	B	L
I	O	
I	I	
O	I	
O	O	

2-3. Which Boolean expression best represents an OR element?

a. A + B = L   b. A/B = L   c. AB = L   d. A − B = L

2-4. Use a series of arrows to show the flow of current around this complete circuit. Note that one switch has been activated.

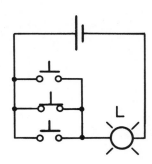

2-5. Write the Boolean expression for this circuit.

_____ = L

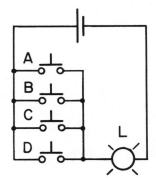

2-6. Complete the following Boolean expressions for an OR.

    a. $1 + 0 = $ _____   b. $1 + 1 = $ _____   c. $0 + 0 = $ _____

2-7. Place a 1 or 0 in each blank to complete the following Boolean expressions. If a missing number may be either a 1 or 0, show a 0.

    a. $0 + $ _____ $= 1$ b. $1 + $ _____ $= 1$ c. $0 + $ _____ $= 0$

Chapter 2
LOGIC ELEMENTS

Name _____

Date _____ Score _____

Instructor _____

SWITCH-BASED NOT ELEMENTS

3-1. Use row a to indicate which switches are NORMALLY OPEN (NO) and which are NORMALLY CLOSED (NC). Use row b to indicate which switches have been shown in the UNACTIVATED (U) position and which are shown in the ACTIVATED (A) position.

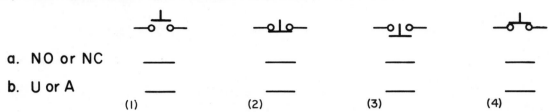

a. NO or NC   ____   ____   ____   ____

b. U or A   ____   ____   ____   ____

      (1)   (2)   (3)   (4)

3-2. For each switch position at the right, indicate the state of the lamp. If lit, L = 1; if out, L = 0.

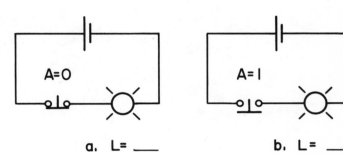

a. L = ____   b. L = ____

3-3. Complete this truth table for a switch-based NOT.

INPUT	OUTPUT
A	L
1	
0	

3-4. Write the Boolean expression that represents the truth table in Problem 3.

_____ = L

3-5. Indicate the values of A, B, and C required to light the lamp in this circuit.

a. A = _____

b. B = _____

c. C = _____

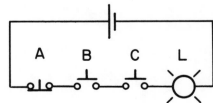

3-6. In this circuit, A = 0. Indicate the values
of B and C necessary to light the lamp.

   a. B = _____

   b. C = _____

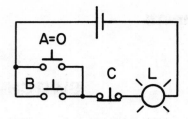

Chapter 2
LOGIC ELEMENTS

Name _____

Date _____ Score _____

Instructor _____

SWITCH-BASED LOGIC CIRCUITS

4-1. Mark each of the following AND, OR, or NOT to indicate the logic element represented.

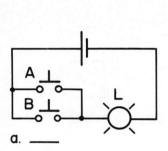

a. _____   b. _____   c. _____

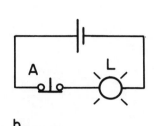

INPUTS		OUTPUT
A	B	L
0	0	0
0	1	1
1	0	1
1	1	1

4-2. Complete this truth table for an AND. Rows are not in standard order.

INPUTS		OUTPUT
A	B	L
0	1	
		1
1		0
0	0	

4-3. For the inputs shown, determine the output of this circuit.
    a. L = 1   b. L = 0

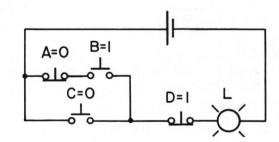

4-4. Complete this truth table for the circuit shown.

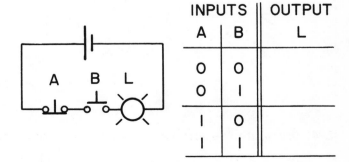

INPUTS		OUTPUT
A	B	L
0	0	
0	1	
1	0	
1	1	

4-5. Mark each of these Boolean expressions AND, OR, or NOT.

    a. _____ A + B + C + D = L   b. _____ $\bar{B}$ = L   c. _____ ABC = L

Chapter 2

LOGIC ELEMENTS

Name _____

Date _____ Score _____

Instructor _____

INTEGRATED CIRCUIT AND

5-1. Redraw this circuit in the following space. If available, use a template. Use either of the following symbol sizes.

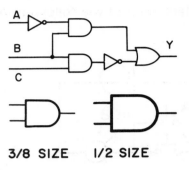

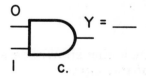

3/8 SIZE   1/2 SIZE

5-2. For each input signal set, determine the logic level at Y.

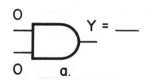 a.

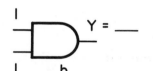 b.

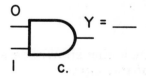 c.

5-3. Complete this switch equivalent of an IC AND by adding the blade to switch y. Place that blade so Y will correspond to the input signals shown.

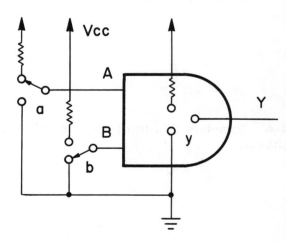

5-4. Evaluate (determine the value of Y) each of these Boolean statements.

a. $1 \times 0 =$ _____   b. $1 \times 1 =$ _____   c. $0 \times 0 =$ _____

5-5. Place a 1 or 0 in each blank to complete the following Boolean expressions. If a missing number may be either 1 or 0, show a 0.

a. ___ $\times 1 = 0$ b. $1 \times$ ___ $= 0$ c. ___ $\times 1 = 1$ d. ___ $\times$ ___ $= 1$

**Chapter 2
LOGIC ELEMENTS**

Name _____

Date _____ Score _____

Instructor _____

INTEGRATED-CIRCUIT AND (7408 TTL or 4081 CMOS)

2-1. Laboratory Activity. Experimentally test the four ANDs in a 7408. Record pin numbers on the symbols to tie your measured results to specific ANDs.

7408

MEASURED OUTPUTS

INPUTS		PREDICTED OUTPUT Y
A	B	
O	O	
O	I	
I	O	
I	I	

2-2. Laboratory Activity. If a 3-input AND is not available, two 2-input ANDs can be used. See the circuit below. The Boolean expression for this circuit is:

$$(AB)C = Y$$

The brackets indicate that A is ANDed with B before the ANDing with C.

Predict the output of this circuit for all possible input sets. Then build the circuit and test your predictions. Place pin numbers on the logic diagram to indicate the logic elements within the 7408 used to construct the circuit.

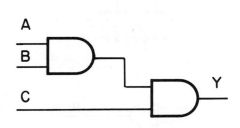

INPUTS			PREDICTED OUTPUT Y	MEAS. OUTPUT Y
A	B	C		
O	O	O		
O	O	I		
O	I	O		
O	I	I		
I	O	O		
I	O	I		
I	I	O		
I	I	I		

25

Chapter 2
LOGIC ELEMENTS

Name _____

Date _____ Score _____

Instructor _____

WIRING OF LOGIC ELEMENTS

7-1. Based on the circuit at the left, place pin numbers on this logic diagram. The chip is a 7408

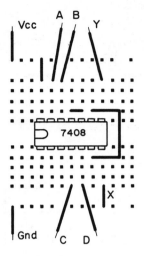

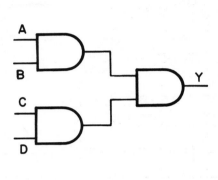

7-2. Refer to lead X in the above circuit. What is its purpose?

   a. Signal input.   b. Auxiliary input.   c. Ground.   d. Chip enable.

7-3. The above circuit uses only three of the four ANDs in the 7408. What action should be taken with respect to the unused AND?
   a. None. It can be ignored.
   b. At least one of its inputs should be connected to an active input.
   c. Unused inputs should be grounded.
   d. Another chip (one with only three ANDs) should be used.

7-4. This 24-pin DIP is being viewed from above. Draw circles around pins 4 and 14.

7-5. This is the foil side of a printed circuit (the side opposite the chips). The bottom of the DIP is shown in dashed lines. Draw circles around pins 1 and 4.

7-6. T   F   Power and ground connections are not shown on this logic diagram. Therefore, ground and power connections need not be made when the circuit is built. Circle the correct answer.

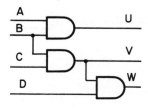

Chapter 2
LOGIC ELEMENTS

Name _____

Date _____ Score _____

Instructor _____

AND ELEMENT TIMING DIAGRAMS

2-1. Construct the output signal of this AND.

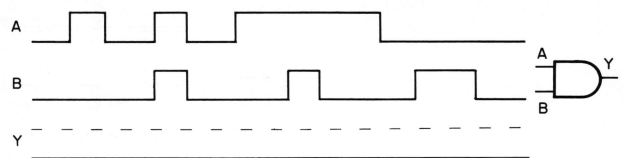

2-2. Repeat Problem 1 for this set of input signals.

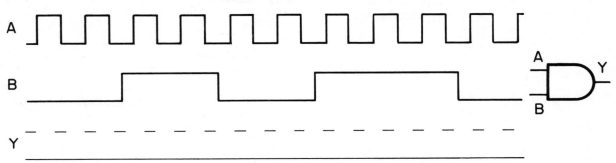

2-3. Repeat Problem 1 for this 3-input AND.

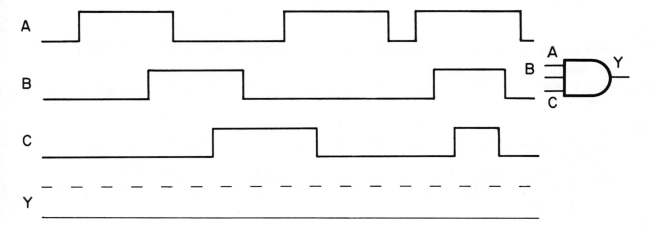

Chapter 2
LOGIC ELEMENTS

Name _____

Date _____ Score _____

Instructor _____

INTEGRATED CIRCUIT OR

9-1. Complete this truth table for a 2-input OR. Rows are not in standard order.

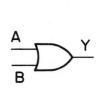

INPUTS		OUTPUT
A	B	Y
0	I	
	O	O
	O	I
I	I	

9-2. Complete this switch equivalent of an IC OR by adding the blade to the switch at its output. Place the blade so Y will correspond to the input signals shown.

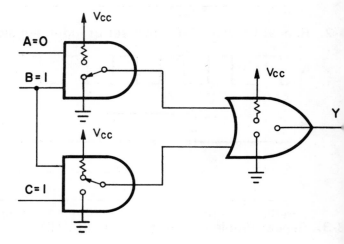

9-3. How many power supplies would probably be used to power the circuit in Problem 2?

    a. None.   b. 1.   c. 2.   d. 3.

9-4. If TTL chips are used in the construction of the circuit in Problem 2, which of the following voltages will be used for Vcc?

    a. 5 V.   b. 18 V.   c. 120 V.

9-5. Which statement best describes the relationship between the ground symbols in the circuit in Problem 2?
    a. They are connected to each other even though no lead is shown.
    b. This circuit contains three independent grounds. They are not interconnected.
    c. Each Vcc has its own ground.

9-6. For each input signal set at the right, determine the value of Y for the circuit in Problem 2.

	A	B	C	Y
a.	0	0	1	_____
b.	1	0	1	_____
c.	1	0	0	_____
d.	1	1	0	_____
e.	0	0	0	_____

Chapter 2

LOGIC ELEMENTS

Name _____

Date _____ Score _____

Instructor _____

INTEGRATED CIRCUIT OR (7432)

O-1. Laboratory Activity. Experimentally test the four ORs in a 7432. Record pin numbers on the symbols to tie your measured results to specific ORs.

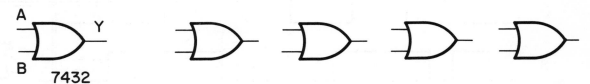

7432

INPUTS		PREDICTED OUTPUT Y
A	B	
O	O	
O	I	
I	O	
I	I	

MEASURED OUTPUTS

Y	Y	Y	Y

O-2. Laboratory Activity. The Boolean expression for the circuit below is:

$$(A + B) + C = Y$$

Predict the output of this circuit for all possible input sets. Then build the circuit and test your predictions.

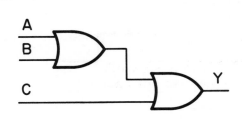

INPUTS			PREDICTED OUTPUT Y	MEAS. OUTPUT Y
A	B	C		
O	O	O		
O	O	I		
O	I	O		
O	I	I		
I	O	O		
I	O	I		
I	I	O		
I	I	I		

Chapter 2
LOGIC ELEMENTS

Name _____

Date _____ Score _____

Instructor _____

OR ELEMENT TIMING DIAGRAMS

11-1. Construct the output signal for this OR element.

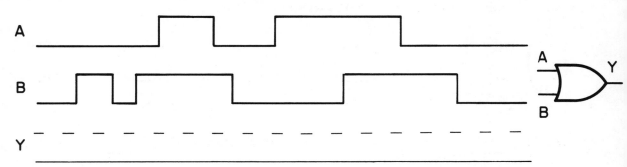

11-2. Repeat Problem 1 for this set of input signals.

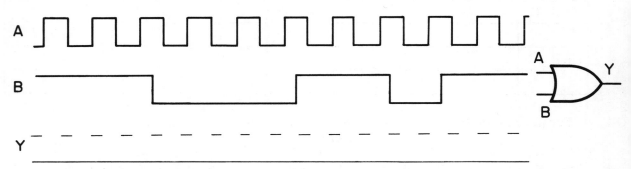

11-3. Repeat Problem 1 for this 3-input OR element.

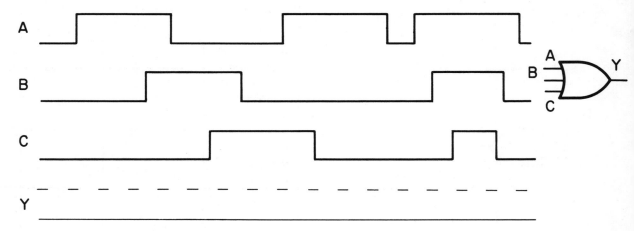

Chapter 2
LOGIC ELEMENTS

Name _____

Date _____ Score _____

Instructor _____

TIMING DIAGRAMS (7408, 7432)

2-1. Laboratory Activity. For the inputs shown, predict the output of this AND element. Record your prediction on the third line of the timing diagram.

Use a 7408 to experimentally verify your prediction. Use the switches on the trainer to create the indicated inputs. Record your results on the fourth line of the timing diagram.

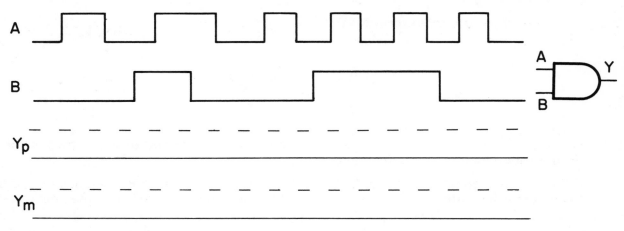

2-2. Laboratory Activity. Repeat Experiment 1 for this 3-input OR element. Use a 7432.

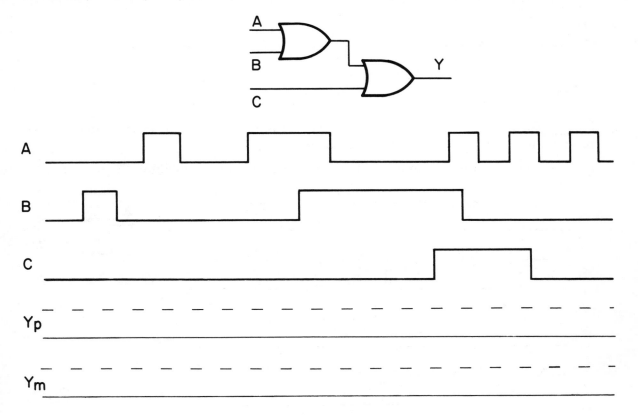

Chapter 2
LOGIC ELEMENTS

Name _____

Date _____ Score _____

Instructor _____

INTEGRATED CIRCUIT NOT (7404, 7432)

13-1. Which is the proper symbol for an IC NOT?

a. A.   b. B.   c. Either.

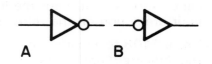

A   B

13-2. Complete this truth table for a NOT.

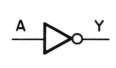

INPUT	OUTPUT
A	Y
0	
	0

13-3. Laboratory Activity. For the input shown, record your prediction of the output of this NOT on the second line of the timing diagram.

If this is an experiment, use a 7404 to test your prediction. Use the switches on the trainer to create the indicated input signal. Record your results on the third line of the timing diagram.

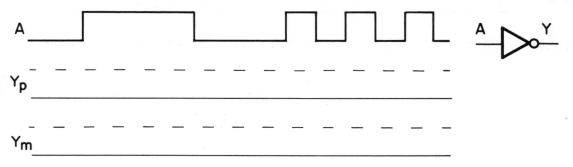

13-4. Laboratory Activity. For the inputs shown, determine the output of this circuit. Record your predictions on third line of the timing diagram.

If this is an experiment, construct the circuit and test your prediction. Record your results on the fourth line.

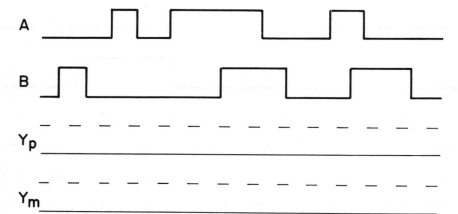

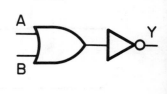

Chapter 2
LOGIC ELEMENTS

Name _____

Date _____ Score _____

Instructor _____

TRUTH TABLE CIRCUIT ANALYSIS (7404, 7408, 7432)

14-1. **Laboratory Activity.** For the input signals shown (solid dot = 1, open dot = 0), determine the logic levels at all points in this circuit. Mark your predictions on the drawing.

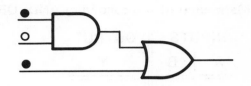

If this is an experiment, build this circuit and test your predictions. Place your results on the drawing and mark them m for measured.

14-2. **Laboratory Activity.** Repeat Problem 1 for this circuit and input signal set.

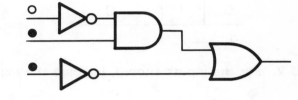

14-3. **Laboratory Activity.** Repeat Problem 1 for this circuit and input signal set.

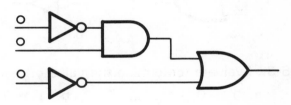

14-4. **Laboratory Activity.** Repeat Problem 1 for this circuit and input signal set.

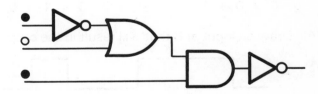

14-5. **Laboratory Activity.** Repeat Problem 1 for this circuit and input signal set.

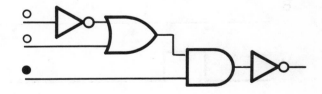

Chapter 2
LOGIC ELEMENTS

Name _____

Date _____ Score _____

Instructor _____

REVIEW OF LOGIC ELEMENTS

15-1. Mark each of the following AND, OR, or NOT to indicate the logic element represented.

INPUTS		OUTPUT
A	B	Y
0	0	0
1	0	0
1	1	1
0	1	0

a. ____

b. ____

ABCD = Y

d. ____

c. ____

$\overline{A}$ = Y

e. ____

15-2. For the inputs shown, indicate the expected outputs.

a.

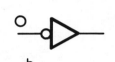

b.

c.

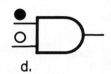

d.

15-3. Evaluate (determine the value of) each of these expressions.

a. $1 + 0$ = _____   b. 1×0 = _____   c. $1 + 1$ = _____

d. $\overline{1}$ = _____   e. $\overline{1} + 0$ = _____   f. $\overline{1} \times 1$ = _____

g. $0 + 0 + \overline{1}$ = _____   h. $\overline{1} + \overline{1} + \overline{0}$ = _____   i. $1 \times 1 \times 1$ = _____

15-4. Draw an input at B that will result in the output shown. Have B at 1 for the shortest possible time.

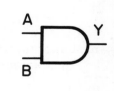

Name _____

Date _____ Score _____

Instructor _____

TRUTH TABLE CIRCUIT ANALYSIS (7404, 7408, 7432)

16-1. Laboratory Activity. For the input signals shown, determine the logic levels at all points in this circuit. Mark your predictions on the drawing.

If this is an experiment, build this circuit and test your predictions. Place your results on the drawing and mark the locations m for measured.

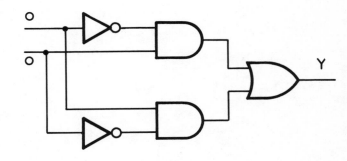

16-2 Laboratory Activity. Repeat Problem 1 for this circuit and input signal set.

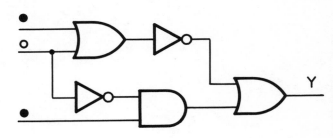

16-3. Laboratory Activity. Repeat Problem 1 for this circuit and input signal set. Note that this circuit has two outputs.

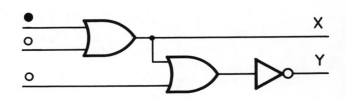

16-4. Laboratory Activity. Repeat Problem 1 for this circuit and input signal set.

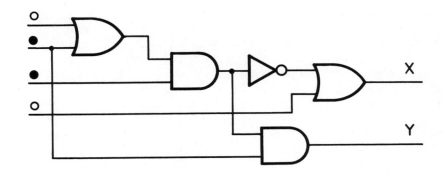

Chapter 3

INTRODUCTION TO TROUBLESHOOTING

Name _____

Date _____ Score _____

Instructor _____

CLASSIFICATION OF FAULTS

1-1. Two circuits were designed to meet the needs of a given application. For the inputs shown, this circuit should output a 1. Which circuit would result in a design fault if it were used?
a. A.
b. B.
c. Either.

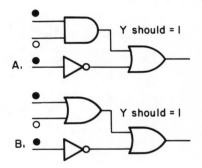

1-2. This circuit contains a fault. The indicated measurements were made at different times. Which best describes this fault?
a. Intermittent.
b. Permanent.

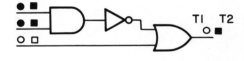

1-3. For the inputs shown, the output of this circuit should be 1. Upon inspection of the printed circuit, a solder bridge was found. Which best describes this fault?
a. External.
b. Internal.

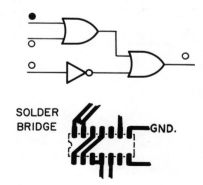

1-4. Voltage measurements were made at the output of this AND. The results are shown at the right. Which best describes this fault?
   a. Parametric.
   b. Logic.

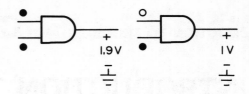

1-5. The output of this AND should be 1 only between times 3 and 4. However, a glitch appears at time 1. Which of the following best describes this fault?
   a. Static.
   b. Dynamic.

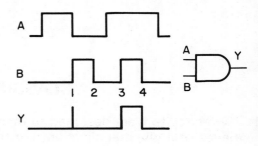

Chapter 3
INTRODUCTION TO TROUBLESHOOTING

Name _____

Date _____ Score _____

Instructor _____

CLASSIFICATION OF FAULTS

The pairs of faults at the right will be used to answer the questions on this sheet.

1a. Design.	1b. Circuit.
2a. Intermittent.	2b. Permanent.
3a. External.	3b. Internal.
4a. Parametric.	4b. Logic.
5a. Static.	5b. Dynamic.

2-1. Below 25 °C, this circuit functions properly. Above that temperature, however, the output of the AND shorts to ground. Classify this fault by circling one letter after each number. The numbers refer to the above list.

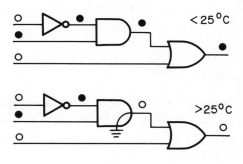

1.   a   b

2.   a   b

3.   a   b

4.   a   b

5.   a   b

2-2. This circuit contains a wiring error. No connection was made between the AND and OR. Classify this fault.

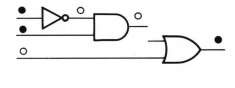

1.   a   b

2.   a   b

3.   a   b

4.   a   b

5.   a   b

2-3. When measured, the output of the upper NOT was found to be 0.9 volts rather than the expected 0.2 volts. Classify this fault.

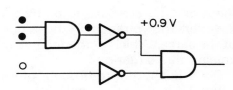

1.   a   b

2.   a   b

3.   a   b

4.   a   b

5.   a   b

Chapter 3
INTRODUCTION TO TROUBLESHOOTING

Name _____

Date _____ Score _____

Instructor _____

TTL OUTPUT SIGNAL VOLTAGES

3-1. This graph shows critical voltages that might be present at the output of the TTL NOT in the following circuit. Complete the graph by placing the proper voltages in the five blanks.

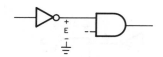

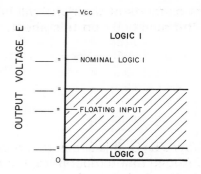

3-2. Match the following measured voltages with the types of faults at the right.

A. Open input lead.
B. Internal short to Vcc.
C. Internal short to ground.
D. Deep internal fault.
E. Parametric fault.

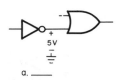

a. ____

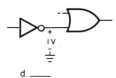

b. ____

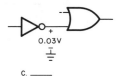

c. ____

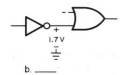

d. ____

3-3. Pins 8 and 13 are connected by a wire. However, the voltages on these pins are not the same. Which of the following best explains this situation?
a. Pin 8 is shorted to ground internally.
b. Pin 13 is open internally.
c. The wire or socket is open.

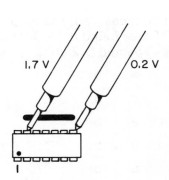

Chapter 3
INTRODUCTION TO TROUBLESHOOTING

Name _____

Date _____ Score _____

Instructor _____

TTL SIGNAL VOLTAGES (7404, VOLTMETER)

4-1. Laboratory Activity. On the line provided below, indicate the nominal (typical or normal range) voltage of a logic 1 at the output of a TTL NOT. Then build the circuit at the right and experimentally determine this voltage. Draw a horizontal line on the graph to represent this voltage and label it ''Typical logic 1.''

Nominal voltage for logic 1 = _____ V

Measured logic 1 voltage = _____ V

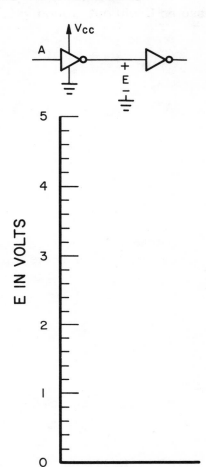

4-2. Laboratory Activity. Change A to 1 and repeat the experiment in 4-1 Laboratory Activity for a logic 0. Plot the voltage on the graph and label it ''Typical logic 0.''

Nominal voltage for logic 0 = _____ V

Measured logic 0 voltage = _____ V

4-3. Laboratory Activity. In the space provided, indicate the expected voltage between a floating input lead and ground. Then open the lead between the two NOTs and measure the voltage shown. Plot this voltage on the graph and label it ''Floating input.''

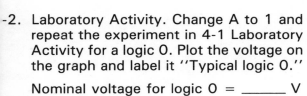

Expected voltage = _____ V

Measured voltage = _____ V

4-4. Laboratory Activity. Reconnect the lead between the NOTs. Set A to 1 and again measure E for a logic 0. Then remove the chip's ground lead as shown. Again measure E. Is this a valid 0?

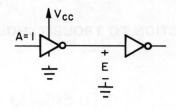

Measured E for valid logic 0 = _____ V

Measured E without ground = _____ V

**Chapter 3
INTRODUCTION TO TROUBLESHOOTING**

Name _____

Date _____ Score _____

Instructor _____

FAULT OBSERVABILITY

5-1. The circuit at the right contains a fault. For which set of inputs does that fault result in an incorrect output? (As an aside, this is the input-signal set that would be used to troubleshoot this circuit.)
   a. A.
   b. B.
   c. C.
   d. D.

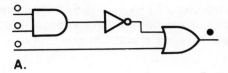

A.

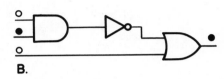

B.

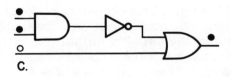

C.

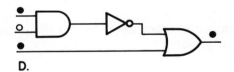

D.

5-2. Repeat Problem 5-1 for the circuit and input sets at the right.
   a. A.
   b. B.
   c. C.

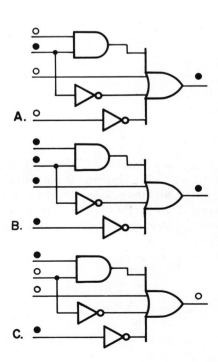

Chapter 3
INTRODUCTION TO TROUBLESHOOTING

Name _____

Date _____ Score _____

Instructor _____

FAULT OBSERVABILITY (7404, 7408, 7432)

6-1. Laboratory Activity. Predict the output of this circuit for each input set in this truth table. Record your predictions in the appropriate column. Then, build the circuit and test your predictions. Record your results in the column marked ''Meas.''

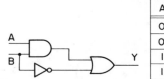

INPUTS		PREDICTED	MEAS.	WITH FAULT
A	B	Y	Y	Y
O	O			
O	I			
I	O			
I	I			

6-2. Laboratory Activity. Disconnect the lead between the NOT and OR in the circuit you have just built. Reconnect as shown at the right. This simulates a short to Vcc fault. For each input set, measure and record the circuit's output. Record each measurement in the column labeled ''With Fault.'' Then, circle the input set that would probably be used to localize this fault.

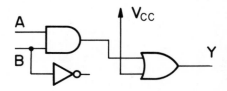

6-3. Laboratory Activity. Repeat the experiment in 6-1 Laboratory Activity for this circuit and partial truth table.

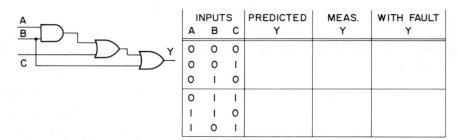

INPUTS			PREDICTED	MEAS.	WITH FAULT
A	B	C	Y	Y	Y
0	0	0			
0	0	I			
0	I	0			
0	I	I			
I	I	0			
I	0	I			

6-4. Laboratory Activity. Repeat the experiment in 6-2 Laboratory Activity for the circuit you have just constructed in 6-3 Laboratory Activity. This time, reconnect the lead between the AND and first OR as shown.

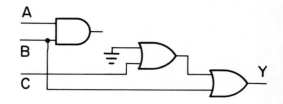

Chapter 3
INTRODUCTION TO TROUBLESHOOTING

Name _____

Date _____ Score _____

Instructor _____

FAULT LOCALIZATION

7-1. There is a fault in this circuit. Localize it by circling the lead where the incorrect signal first appears.

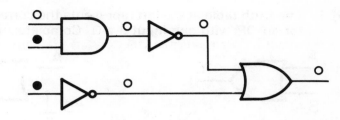

7-2. Repeat Problem 7-1 for this circuit and input signal set.

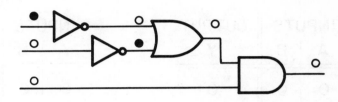

7-3. Repeat Problem 7-1 for this circuit and input signal set.

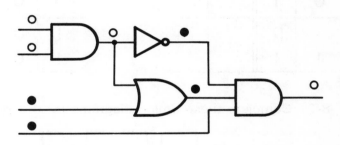

7-4. Repeat Problem 7-1 for this circuit and input signal set.

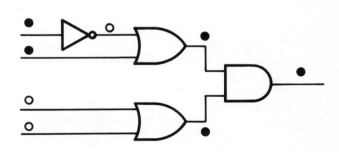

7-5. Repeat Problem 7-1 for this circuit and input signal set.

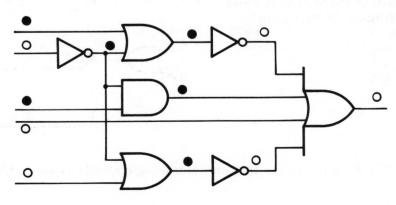

Chapter 3
INTRODUCTION TO TROUBLESHOOTING

Name _____

Date _____ Score _____

Instructor _____

FAULTS OF THE "STUCK-AT" TYPE

8-1. The truth table at the left represents the correct operation of an OR. Complete the second tab for an OR with one input s-a-0. Complete the third table for an OR with one input s-a-1.

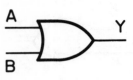

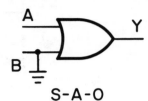

S-A-0

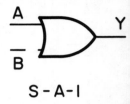

S-A-I

INPUTS		OUTPUT
A	B	Y
0	0	0
0	1	1
1	0	1
1	1	1

INPUTS		OUTPUT
A	B	Y
0	0	
0	0	
1	0	
1	0	

INPUTS		OUTPU
A	B	Y
0	1	
0	1	
1	1	
1	1	

8-2. Complete these timing diagrams for the indicated stuck-at faults.

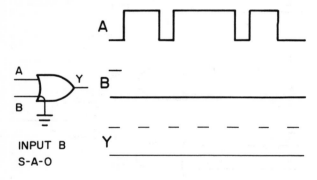

INPUT B
S-A-0

INPUT B
FLOATING
S-A-I

8-3. This OR contains a fault. Based on the timing diagram, which of the following is the most likely description of that fault?
  a. Output is s-a-0.
  b. Input A is s-a-0.
  c. Input A is s-a-1.
  d. Input B is s-a-0.
  e. Input B is s-a-1.

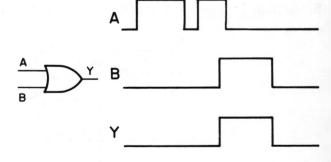

8-4. T   F   When TTL logic family is used, inputs treat open leads as s-a-1. Circle correct answe

Name _____

Date _____ Score _____

Instructor _____

FAULTS OF THE "STUCK-AT" TYPE

-1. The truth table at the left represents the correct operation of an AND. Complete the second table for an AND with one input s-a-0. Complete the third table for an AND with one input s-a-1.

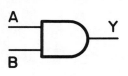 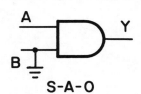

S-A-O   S-A-I

INPUTS		OUTPUT
A	B	Y
O	O	O
O	I	O
I	O	O
I	I	I

INPUTS		OUTPUT
A	B	Y
O	O	
O	O	
I	O	
I	O	

INPUTS		OUTPUT
A	B	Y
O	I	
O	I	
I	I	
I	I	

-2. Complete these timing diagrams for the indicated stuck-at faults.

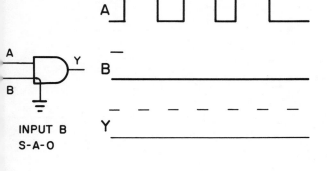

INPUT B
S-A-O

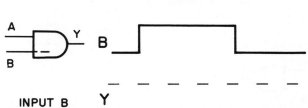

INPUT B
FLOATING
S-A-I

-3. This AND contains a fault. Based on the timing diagram, which of the following is the most likely description of that fault?
   a. Output is s-a-1.
   b. Input A is s-a-0.
   c. Input A is s-a-1.
   d. Input B is s-a-0.
   e. Input B is s-a-1.

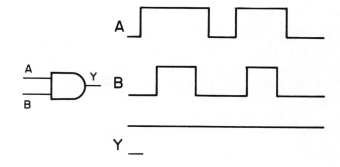

9-4. If the input lead to a NOT floats, which of the following best describes its output signal? Assum
     a TTL NOT is used.
     a. s-a-1.
     b. s-a-0.
     c. It will output the signal applied to its input, but it will not be inverted.

Name _____

Date _____ Score _____

Instructor _____

FAULTS OF THE "STUCK-AT" TYPE (7408, 7432)

10-1. Laboratory Activity. Assemble the circuit shown. With the rest of the laboratory team away from the bench, have a member of your team make one of the indicated connections between points D and E. Then cover this portion of the circuit with a piece of paper.

The team is to electrically determine which connection was made. This is to be accomplished by applying various signal sets to inputs A, B, and C and observing lamps L1, L2, and L3. Record your prediction by circling a, b, or c below. Then remove the paper to determine if the prediction is correct.

Predicted connection: a,   b,   c.   (Correct, Incorrect)

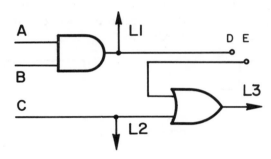

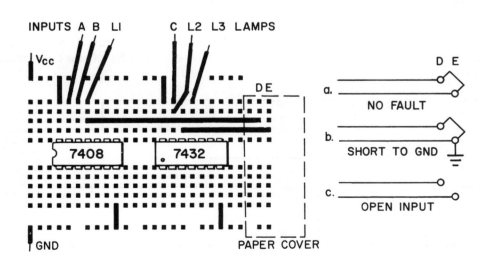

10-2. Laboratory Activity. Repeat the experiment in 10-1 Laboratory Activity. Have a different member of your team make the connection.

Predicted connection: a,   b,   c.   (Correct, Incorrect)

Chapter 3
INTRODUCTION TO TROUBLESHOOTING

Name _____

Date _____ Score _____

Instructor _____

DYNAMIC FAULTS

11-1. The output of this AND should be 0 at all times, since A and B should never be 1 at the same time. However, B has been offset as shown. Plot the output of this AND. (In most race situations, offsets are very small, and special instruments are needed to detect the resulting glitches.)

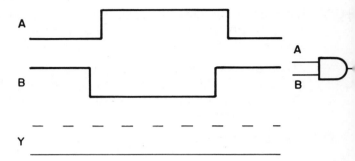

11-2. The output of the following AND should always be 0; the output of the OR should always be 1. In each case, however, there are two race conditions. Add glitches to these diagrams to indicate the locations of the race conditions (two per timing diagram). The glitches are negative going at the output of the OR.

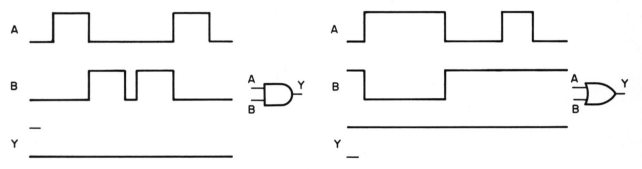

11-3. The signal at the output of this AND gate is applied to a counter. That counter should count the three obvious pulses. It counts five. Indicate the locations of the two glitches that cause this error.

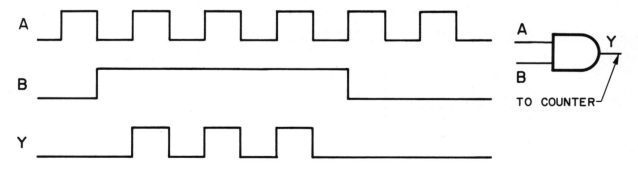

11-4. In general, which best describes faults caused by race conditions and the resulting glitches?
   a. Design error.
   b. Circuit fault.

11-5. T F Replacing faulty chips is the usual approach to reparing faults caused by race conditions. Circle the correct answer.

Chapter 3
INTRODUCTION TO TROUBLESHOOTING

Name _____

Date _____ Score _____

Instructor _____

DYNAMIC MEASUREMENTS

12-1. Oscilloscopes and logic analyzers can be used to do troubleshooting. The following OR is faulty. Circle the first place in this timing diagram where the symptom of the fault can be detected.

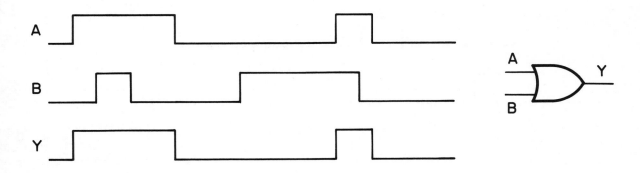

12-2. Repeat Problem 12-1 for this AND.

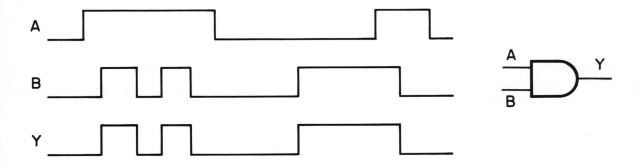

12-3. Repeat Problem 12-1 for this NOT and AND.

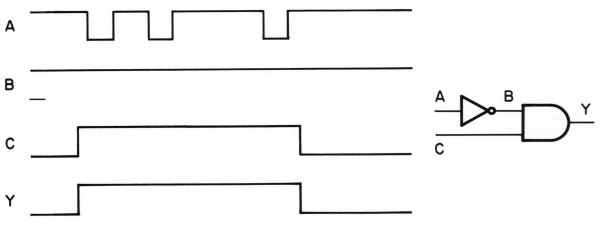

Chapter 3
INTRODUCTION TO TROUBLESHOOTING

Name _____

Date _____ Score _____

Instructor _____

DYNAMIC MEASUREMENTS

13-1. Complete this timing diagram for the circuit shown.

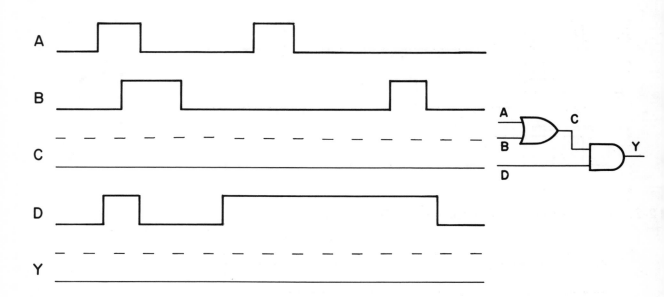

13-2. This timing diagram might be displayed on an oscilloscope or logic analyzer. One of the element
is faulty. Circle the first point on the diagram where this fault becomes apparent.

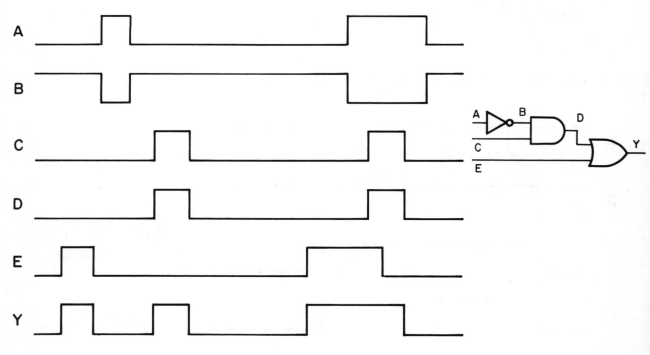

Chapter 3
INTRODUCTION TO TROUBLESHOOTING

Name _____

Date _____ Score _____

Instructor _____

WIRING ERRORS

14-1. Wiring errors are external faults. They can often be isolated through visual inspections. This circuit contains a common wiring error. Use heavy lines to indicate the corrective action that would be taken.

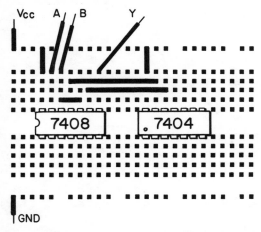

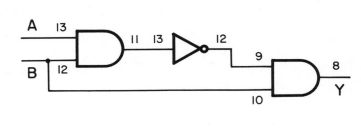

14-2. Repeat Problem 14-1 for this circuit.

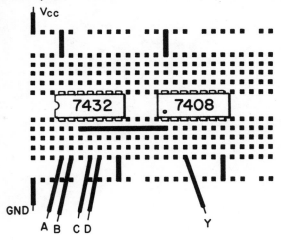

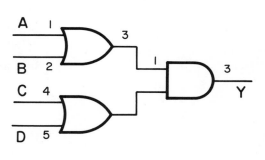

14-3. Repeat Problem 14-1 for this circuit.

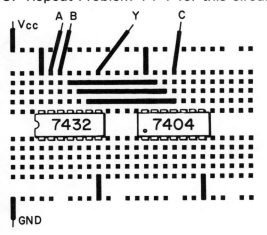

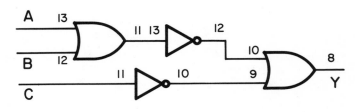

Chapter 4

COMBINATIONAL LOGIC—AND, OR, NOT

Name _____

Date _____ Score _____

Instructor _____

TRANSFORMATION 1: LOGIC DIAGRAM TO TRUTH TABLE
(7404, 7408, 7432)

1-1. Laboratory Activity. Use truth table circuit analysis to complete this truth table for the following circuit. Record your predictions in the column marked Yp. If this is an experiment, build the circuit and test your predictions. Record your results in the column marked Ym.

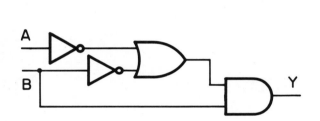

INPUTS		OUTPUT	OUTPUT
A	B	Y_p	Y_m
0	0		
0	1		
1	0		
1	1		

1-2. Laboratory Activity. Repeat Problem 1 for this circuit and partial truth table.

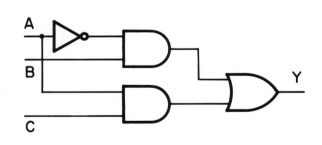

INPUTS			OUTPUT	OUTPUT
A	B	C	Y_p	Y_m
0	0	0		
0	1	0		
1	1	0		
0	0	1		

1-3. Laboratory Activity. Repeat Problem 1 for the circuit with two outputs and a partial truth table.

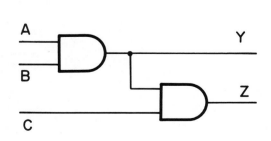

INPUTS			OUTPUTS		OUTPUTS	
A	B	C	Y_p	Z_p	Y_m	Z_m
1	1	1				
0	1	0				
1	1	0				
0	1	1				

1-4. Laboratory Activity. Repeat Problem 1 for this circuit and partial truth table.

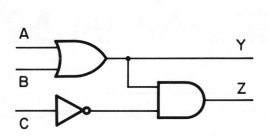

INPUTS			OUTPUTS		OUTPUTS	
A	B	C	Y_p	Z_p	Y_m	Z_m
1	1	1				
0	1	0				
1	1	0				
0	1	1				

Chapter 4
COMBINATIONAL LOGIC—AND, OR, NOT

Name _____

Date _____ Score _____

Instructor _____

TRANSFORMATION 2: BOOLEAN EXPRESSION TO TRUTH TABLE

4-1. Based on the following expression, complete this truth table.

$A\overline{B} + A = Y$

INPUTS		OUTPUT
A	B	Y
O	O	
O	I	
I	O	
I	I	

4-2. Repeat Problem 1 for this expression and partial truth table.

$B(A + \overline{C}) = Y$

INPUTS			OUTPUT
A	B	C	Y
I	I	I	
O	O	O	
O	I	O	
I	I	O	

4-3. Repeat Problem 1 for this expression and partial truth table.

$AB(C + A)(C + B) = Y$

INPUTS			OUTPUT
A	B	C	Y
I	I	O	
O	I	I	
I	O	I	
I	I	I	

4-4. Repeat problem 1 for these expressions giving outputs at Y and Z and with a partial truth table.

$A + B\overline{C} = Y$

$\overline{A} + \overline{B}C = Z$

INPUTS			OUTPUTS	
A	B	C	Y	Z
I	O	O		
O	I	O		
O	O	I		
I	I	O		

Chapter 4
COMBINATIONAL LOGIC—AND, OR, NOT

Name _____

Date _____ Score _____

Instructor _____

TRANSFORMATION 2: BOOLEAN EXPRESSION TO TRUTH TABLE

3-1. Complete this partial truth table based on the following expression.

$$\overline{AB} + \overline{C} = Y$$

INPUTS			OUTPUT
A	B	C	Y
I	I	I	
O	I	I	
I	O	O	
O	O	I	

3-2. Repeat Problem 1 for these expressions.

$$\overline{AB} = Y$$

$$\overline{A}\,\overline{B} = Y$$

INPUTS		OUTPUTS	
A	B	Y	Z
O	O		
O	I		
I	O		
I	I		

3-3. Repeat Problem 1 for this expression.

$$\overline{A + B + \overline{C}} + C = Y$$

INPUTS			OUTPUT
A	B	C	Y
I	I	I	
I	I	O	
O	O	O	
I	O	O	

3-4. Repeat Problem 1 for this expression.

$$\overline{A + B}\ \overline{AC} = Y$$

INPUTS			OUTPUT
A	B	C	Y
O	O	O	
O	O	I	
I	O	O	
I	I	I	

Name _____

Date _____ Score _____

Instructor _____

TRANSFORMATION 3: LOGIC DIAGRAM TO BOOLEAN EXPRESSION

Write the expression for each of the following logic diagrams. Do not simplify.

4-1. _____ = Y

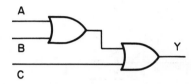

4-2. _____ = Y

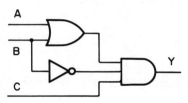

4-3. _____ = Y

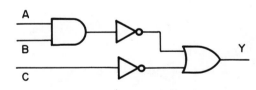

4-4. _____ = Y

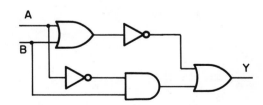

4-5. _____ = Y

4-6. _____ = Y

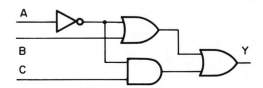

Chapter 4
COMBINATIONAL LOGIC—AND, OR, NOT

Name _____

Date _____ Score _____

Instructor _____

TRANSFORMATION 3: LOGIC DIAGRAM TO BOOLEAN EXPRESSION
(7404, 7408, 7432)

5-1. Laboratory Activity. First, write the expression for this circuit in the space provided. Then, us
your expression to predict the circuit's output for each input set. If this is an experiment, buil
the circuit and test your predictions. If this is not an experiment, test your predictions by apply
ing truth table circuit analysis to the circuit. Record your predictions and test results in the column
provided.

_____ = Y

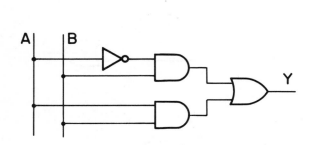

INPUTS		PREDICTED	TEST OF
A	B	Y	Y
O	O		
O	I		
I	O		
I	I		

5-2. Laboratory Activity. Repeat Problem 1 for this circuit.

_____ = Y

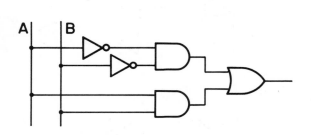

INPUTS		PREDICTED	TEST OF
A	B	Y	Y
O	O		
O	I		
I	O		
I	I		

Chapter 4
COMBINATIONAL LOGIC—AND, OR, NOT

Name _____

Date _____ Score _____

Instructor _____

TRANSFORMATION 3: LOGIC DIAGRAM TO BOOLEAN EXPRESSION
(7404, 7408, 7432)

5-1. Laboratory Activity. First, write the expression for this circuit in the space provided. Then, use the expression to predict the circuit's output for each input set shown. If this is an experiment, build the circuit and test your predictions. If this is not an experiment, test your predictions by applying truth table circuit analysis to the circuit. Record your predictions and results in the columns provided.

_____ = Y

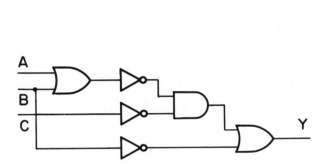

INPUTS			PREDICTED	TEST OF
A	B	C	Y	Y
1	0	0		
0	0	1		

5-2. Laboratory Activity. Repeat Problem 1 for this circuit.

_____ = Y

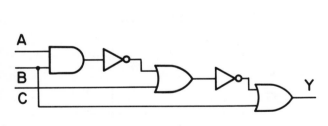

INPUTS			PREDICTED	TEST OF
A	B	C	Y	Y
0	1	0		
1	1	0		
0	0	0		

Chapter 4
COMBINATIONAL LOGIC—AND, OR, NOT

Name _____

Date _____ Score _____

Instructor _____

TRANSFORMATION 4: BOOLEAN EXPRESSION TO LOGIC DIAGRAM

Draw the circuit represented by each expression. Do not simplify. Use the space at the right for you drawings.

7-1.

$$AB + C = Y$$

7-2.

$$AB + \overline{A} = Y$$

7-3.

$$\overline{A}\overline{B} + \overline{C} = Y$$

7-4.

$$\overline{AB} + \overline{C}\overline{D} = Y$$

7-5.

$$\overline{\overline{AB} + C} = Y$$

7-6.

$$\overline{(\overline{A+B})C + AB} = Y$$

hapter 4
OMBINATIONAL LOGIC—AND, OR, NOT

Name _____

Date _____ Score _____

Instructor _____

TRANSFORMATION 4: BOOLEAN EXPRESSION TO LOGIC DIAGRAM
(7404, 7408, 7432)

-1. Laboratory Activity. In the space at the right, draw the logic diagram represented by the following expression. Do not simplify. Also based on the expression, predict the circuit's output for the input signal sets shown in the partial truth table. If this is a laboratory experiment, build the circuit and test your predictions. If this is not an experiment, test your predictions by applying truth table circuit analysis to the circuit. Record your predictions and results in the columns provided.

$$\overline{AB} + C = Y$$

INPUTS			PREDICTED	TEST OF
A	B	C	Y	Y
I	I	I		
O	I	I		
O	I	O		

-2. Laboratory Activity. Repeat Problem 1 for this expression.

$$(\overline{A + B})C + AB = Y$$

INPUTS			PREDICTED	TEST OF
A	B	C	Y	Y
O	O	O		
I	I	O		
I	O	I		

Chapter 4
COMBINATIONAL LOGIC—AND, OR, NOT

Name _____

Date _____ Score _____

Instructor _____

TRANSFORMATION 5: BOOLEAN EXPRESSION FROM TRUTH TABLE
(7404, 7408, 7432)

9-1. Using sum-of-products, write an expression to represent this truth table. Do not simplify.

_____ =

INPUTS		GIVEN	TEST OF
A	B	Y	Y
0	0	1	
0	1	0	
1	0	0	
1	1	1	

9-2. Laboratory Activity. Draw the circuit represented by the above expression. Do not simplify. If this is an experiment, build the circuit and test it by applying the signal sets in the truth table. If this is not an experiment, test your circuit by applying truth table circuit analysis to the circuit diagram. Record your results in the column provided.

9-3. Repeat Problem 1 for this truth table. Again, do not simplify.

_____ =

	INPUTS			GIVEN	TEST OF
	A	B	C	Y	Y
*	0	0	0	0	
	0	0	1	0	
	0	1	0	0	
*	0	1	1	1	
*	1	0	0	0	
*	1	0	1	1	
	1	1	0	0	
	1	1	1	0	

9-4. Laboratory Activity. Repeat Problem 2 for the above expression. Test your circuit for the input signal sets marked * in the truth table.

64

Chapter 4
COMBINATIONAL LOGIC—AND, OR, NOT

Name _____

Date _____ Score _____

Instructor _____

TRANSFORMATION 5: BOOLEAN EXPRESSION FROM TRUTH TABLE
(7404, 7408, 7432)

10-1. Using sum-of-products, write an expression to represent this truth table. Do not simplify.

_____ = Y

INPUTS		GIVEN	TEST OF
A	B	Y	Y
0	0	1	
0	1	0	
1	0	1	
1	1	0	

10-2. Laboratory Activity. Draw the circuit represented by the above expression. Do not simplify. If this is an experiment, build the circuit and test it by applying the signal sets in the truth table. If this is not an experiment, test your circuit by applying truth table circuit analysis to the circuit diagram. Record your results in the column provided.

10-3. Repeat Problem 1 for the above truth table, but this time use product-of-sums. Do not simplify.

_____ = Y

10-4. Laboratory Activity. Repeat Problem 2 for the above expression. Again, do not simplify. Record the test results in the table at the right. In terms of their truth tables, is this circuit electrically equivalent to the one drawn in Problem 2?

INPUTS		TEST OF
A	B	Y
0	0	
0	1	
1	0	
1	1	

Chapter 5

NAND, NOR, XOR ELEMENTS

Name _____

Date _____ Score _____

Instructor _____

DESCRIPTION OF NAND

-1. Complete this truth table to show the
action of a NAND.

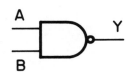

INPUTS		OUTPUT
A	B	Y
O	O	
O	I	
I	O	
I	I	

-2. Show how this AND and NOT would be
connected to produce a NAND logic
element.

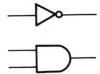

-3. Which of these Boolean expressions best represents the action of a NAND?

a. $\overline{A}\overline{B} = Y$ b. $\overline{AB} = Y$ c. $\overline{A + B} = Y$

-4. Construct the expected output of this NAND based on the indicated input signals.

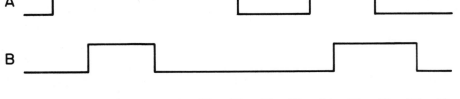

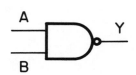

1-5. This timing diagram emphasizes the NAND as a gate. Construct the expected output of the NAND based on the indicated input signals.

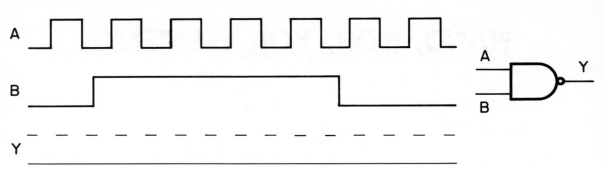

Name _____

Date _____ Score _____

Instructor _____

DE MORGAN'S THEOREM APPLIED TO NAND ELEMENTS

-1. For each input signal set, determine Y and Z. In effect, this problem proves DeMorgan's theorem when applied to NANDs.

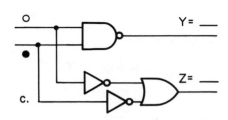

a.

Y= __

Z= __

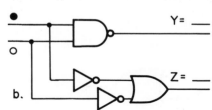

b.

Y= __

Z= __

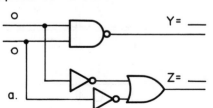

c.

Y= __

Z= __

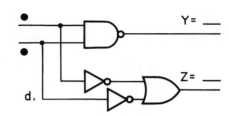

d.

Y= __

Z= __

-2. For the inputs shown, construct the expected output of the NAND. Note the use of the OR-based symbol. You will find that this symbol makes the task easier.

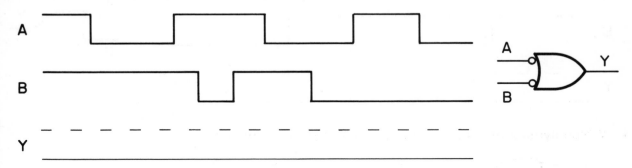

A

B

Y

-3. Indicate the actions of these NAND-based circuits by matching them with the symbols at the right.

a. __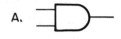

b. __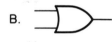

c. __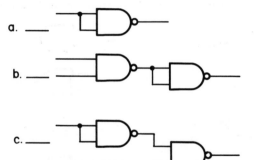

A.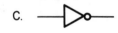

B.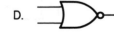

C.

D.

Chapter 5
NAND, NOR, XOR ELEMENTS

Name _____

Date _____ Score _____

Instructor _____

DESCRIPTION OF NOR

3-1. Complete this truth table to show the action of a NOR. Inputs are not in standard order.

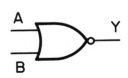

INPUTS		OUTPUT
A	B	Y
I	I	
O	I	
O	O	
I	O	

3-2. Which of the following best represents the action of a NOR?

a. $\overline{A} + \overline{B} = Y$ b. $\overline{AB} = Y$ c. $\overline{A + B} = Y$

3-3. For the indicated input signals, construct the output of this NOR.

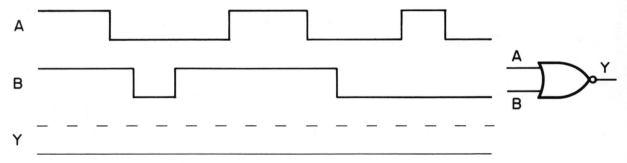

3-4. Which symbol best represents the DeMorgan implementation of a NOR?

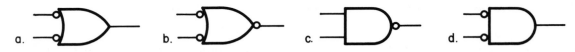

3-5. Indicate the actions of these NOR-based circuits by matching them with the symbols at the right

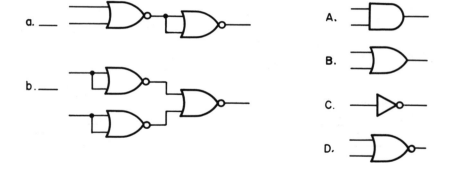

Name _____

Date _____ Score _____

Instructor _____

ACTIVE-HIGH/ACTIVE-LOW

5-1. Input G1 is active-high. Which circuit best describes the signals at A and B necessary to activate G1?

 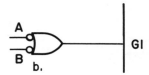

5-2. In this case, C is active-low. Which circuit best describes the signals at A and B necessary to activate C?

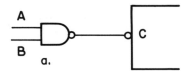 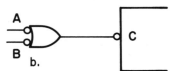

5-3. Redraw this circuit to emphasize its action as an AND. Do this by using DeMorgan implementations for any or all of the NORs.

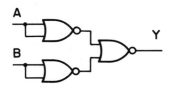 

5-4. The ENABLE on this device is active-high. To convert it to an active-low, a NOT has been placed in its input lead. Which is the preferred symbol for the NOT in this application?

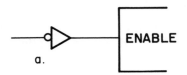 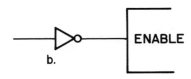

5-5. This device has two outputs, Q and $\bar{Q}$.
$\bar{Q}$ is always the inverse of Q. If bubble
notation were used at the output of this
device, on which lead would the bubble
be placed?
a. On Q.
b. On $\bar{Q}$.
c. On both Q and $\bar{Q}$.

Chapter 5
NAND, NOR, XOR ELEMENTS

Name _____

Date _____ Score _____

Instructor _____

NAND LOGIC ELEMENT (7400)

5-1. Laboratory Activity. Experimentally test the four NANDs in a 7400. Record pin numbers on the following symbols to tie your measured results to specific NANDs within the chip.

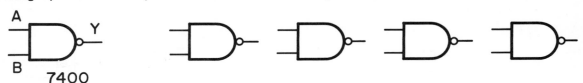

7400

MEASURED OUTPUTS

INPUTS		PREDICTED OUTPUT		Y	Y	Y	Y
A	B	Y					
O	O						
O	I						
I	O						
I	I						

5-2. Laboratory Activity. This NAND functions as a NOT. Predict the output of this circuit and then experimentally test your predictions.

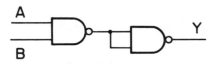

INPUT	PREDICTED	MEAS.
A	Y	Y
O		
I		

5-3. Laboratory Activity. This circuit functions as an AND. Predict the output of this circuit for each input set shown in the truth table and then experimentally test your predictions.

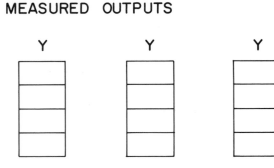

INPUTS		PREDICTED	MEAS.
A	B	Y	Y
O	O		
O	I		
I	O		
I	I		

5-4. Laboratory Activity. This circuit results in an OR. Repeat the experiment in Problem 3 for this circuit.

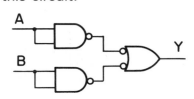

INPUTS		PREDICTED	MEAS.
A	B	Y	Y
O	O		
O	I		
I	O		
I	I		

Chapter 5
NAND, NOR, XOR ELEMENTS

Name _____

Date _____ Score _____

Instructor _____

NOR LOGIC ELEMENT (7402)

5-1. Laboratory Activity. For the input signals shown, predict the output of this NOR. Record your predictions on the graph marked Yp. Then experimentally test your predictions. Graph the measured outputs on Ym. Use the switches on your trainer to obtain the indicated input signals.

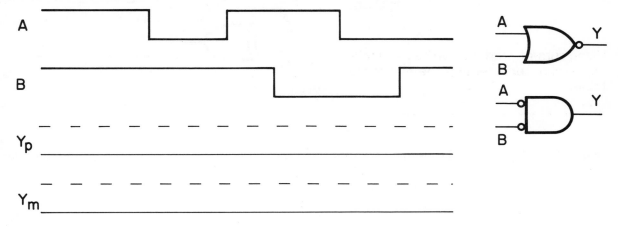

5-2. Laboratory Activity. Connected as shown, a NOR can be used as a NOT. Predict the output of this circuit and then experimentally test your predictions.

INPUT A	PREDICTED Y	MEAS. Y
0		
1		

5-3. Laboratory Activity. This circuit functions as an AND. Predict its output for each input in the truth table. Then experimentally test your predictions.

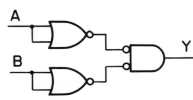

INPUTS A	B	PREDICTED Y	MEAS. Y
0	0		
0	1		
1	0		
1	1		

5-4. Laboratory Activity. This circuit results in an OR. Repeat the experiment in Problem 3 for this circuit.

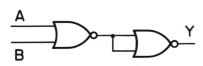

INPUTS A	B	PREDICTED Y	MEAS. Y
0	0		
0	1		
1	0		
1	1		

Chapter 5
NAND, NOR, XOR ELEMENTS

Name _____

Date _____ Score _____

Instructor _____

XOR LOGIC ELEMENTS

7-1. Complete this truth table to show the action of an XOR.

7-2. Using sum-of-products, write the Boolean expression representing the above truth table.

_____ = Y

INPUTS		OUTPUT
A	B	Y
O	O	
O	I	
I	O	
I	I	

7-3. Using AND, OR, and NOT elements, draw the logic diagram representing the above expression

7-4. Based on the following input signals, construct the expected output of this XOR.

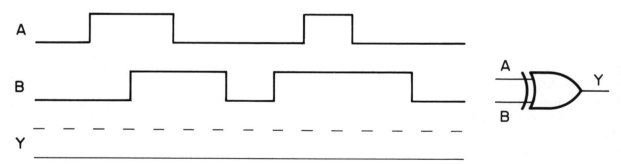

7-5. The following input signals produce a glitch. Find the race condition and draw the glitch on the plot of Y.

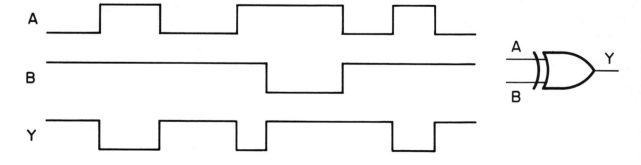

Name _____

Date _____ Score _____

Instructor _____

DESCRIBING LOGIC ELEMENTS

8-1. For each set of input signals, indicate the output signals of these logic elements. The inputs are not in standard order.

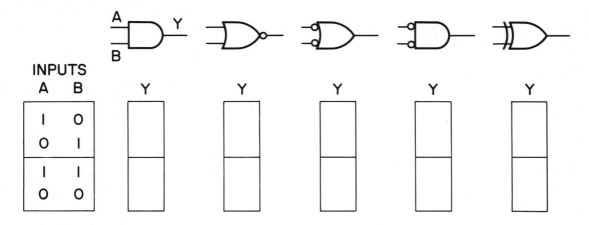

INPUTS

A	B
I	O
O	I
I	I
O	O

8-2. Match these elements with the DeMorgan implementations at the right.

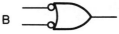

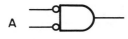

8-3. For the input signals shown, determine the signal at the output of this circuit. A graph has been provided for the intermediate point in the circuit.

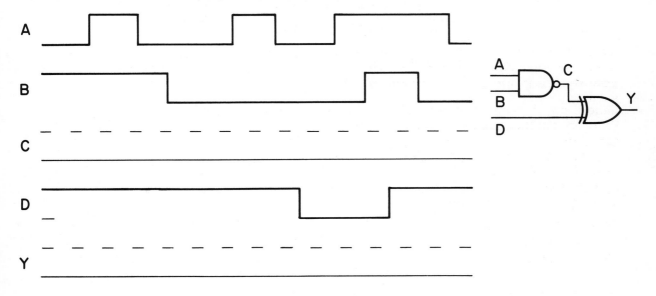

Chapter 5
NAND, NOR, XOR ELEMENTS

Name _____

Date _____ Score _____

Instructor _____

XOR CIRCUITS (7400)

9-1. Complete the column labeled "PRE-DICTED" in this truth table for an XOR.

INPUTS		PREDICTED	5-9-2	5-9-3
A	B	Y	Y	Y
O	O			
O	I			
I	O			
I	I			

9-2. This NAND-based circuit functions as an XOR. Use truth table circuit analysis to determine its output for each input signal set. Record your results in the column marked 5-9-2 in the truth table.

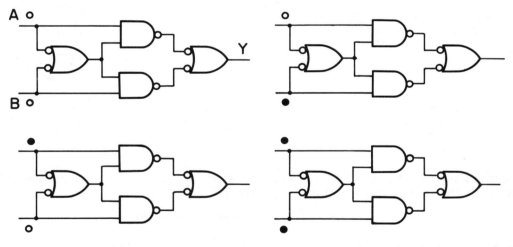

9-3. Laboratory Activity. If this is an experiment, build the above circuit using a 7400. Use one of the above logic diagrams to record pin numbers. Test your predictions experimentally and record your results in the column marked 5-9-3.

9-4. Laboratory Activity. If this is an experiment, use the circuit constructed in the experiment in Problem 3 in the following manner. At the same time, switch A and B back and forth between 0,0 and 1,1. Output Y should remain at 0. See the timing diagram at the right. However, the indicator lamp at Y is likely to light for an instant when the inputs are switched. Explain this fault.

_____

_____

_____

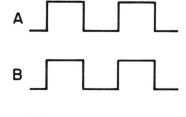

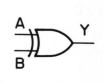

Chapter 5
NAND, NOR, XOR ELEMENTS

Name _____

Date _____ Score _____

Instructor _____

LOGIC CIRCUIT ACTION (7400, 7402, 7404, 7432, 7486)

10-1. Based on the expression at the right, predict the values of Y for the input sets in this partial truth table.

$$\overline{A+B}+\overline{A+C} = Y$$

INPUTS			PREDICTED	TEST OF
A	B	C	Y	Y
I	I	O		
O	I	O		
O	O	I		

10-2. Draw the logic diagram represented by the above expression. Do not simplify. Represent the expression as closely as possible. For example, two NORs and an OR should be used to implement this expression.

10-3. Laboratory Activity. If this is an experiment, construct the above circuit and test its action using the input signal sets in the truth table. If this is not an experiment, use truth table circuit analysis to test your circuit. Record your results in the column marked Test of Y.

10-4. Repeat Problem 1 for this expression.

$$\overline{(\overline{A}B+A\overline{B})C} = Y$$

INPUTS			PREDICTED	TEST OF
A	B	C	Y	Y
O	I	I		
O	I	O		
I	I	I		

10-5. Repeat Problem 2 for the above expression.

10-6. Laboratory Activity. Repeat the experiment in Problem 3 for the above circuit.

77

Chapter 5
NAND, NOR, XOR ELEMENTS

Name _____

Date _____ Score _____

Instructor _____

EXPRESSION FOR CIRCUITS CONTAINING NANDS AND NORS (7400, 7402)

11-1. Mark each of these elements NAND or NOR.

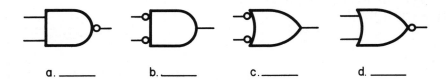

a. _____   b. _____   c. _____   d. _____

11-2. Repeat Problem 1 for the elements in this circuit. Then write the expression for the circuit. Do not simplify.

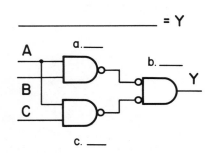

_____ = Y

a. ___
b. ___
c. ___

INPUTS			PREDICTED	TEST OF
A	B	C	Y	Y
I	I	I		
O	I	I		
I	I	O		

11-3. Based on the above expression, predict the circuit's output for each input signal set in the partial truth table.

11-4. Laboratory Activity. If this is an experiment, build the above circuit and test your predictions. If this is not an experiment, apply truth table circuit analysis to the above circuit to test your predictions. Record your results in the column marked Test of Y.

11-5. Repeat Problem 2 for this circuit.

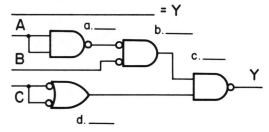

_____ = Y

a. ___
b. ___
c. ___
d. ___

INPUTS			PREDICTED	TEST OF
A	B	C	Y	Y
I	O	O		
O	O	O		
O	I	I		

11-6. Repeat Problem 3 for the signal sets in the above truth table.

11-7. Laboratory Activity. Repeat the experiment in Problem 4 for the circuit and truth table in Problem 5.

Chapter 5
NAND, NOR, XOR ELEMENTS

Name _____

Date _____ Score _____

Instructor _____

NAND AND NOR AS UNIVERSAL LOGIC ELEMENTS (7400, 7402)

2-1. Based on the expression at the right, predict Y for each input signal set in this partial truth table.

$$AC + B = Y$$

INPUTS			PREDICTED	TEST OF
A	B	C	Y	Y
I	I	O		
O	I	O		
O	O	I		

2-2. Laboratory Activity. The following circuit is a NAND implementation of the above expression. Elements 1 and 2 serve to AND A and C. Elements 3, 4, and 5 form an OR. In the middle circuit, redundant NOTs 2 and 3 have been removed. At the right, an OR-based symbol has been used for the output NAND.

If this is an experiment, build the simplified circuit and test its action using the input signal sets in the above truth table. If this is not an experiment, apply truth table circuit analysis to test your predictions. Record your results in the column marked Test of Y.

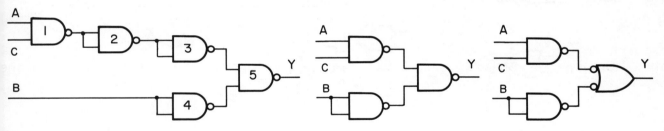

2-3. Repeat Problem 1 for this expression and partial truth table.

$$(A + C)\overline{B} = Y$$

INPUTS			PREDICTED	TEST OF
A	B	C	Y	Y
I	I	I		
I	O	O		
O	O	I		

2-4. Laboratory Activity. Use NORs to implement the above expression. Remove redundant NOTs. Then repeat the experiment in Problem 2 for this NOR-based circuit.

Chapter 5
NAND, NOR, XOR ELEMENTS

Name _____

Date _____ Score _____

Instructor _____

UNIVERSAL LOGIC ELEMENTS

13-1. Complete this table by adding the four missing elements.

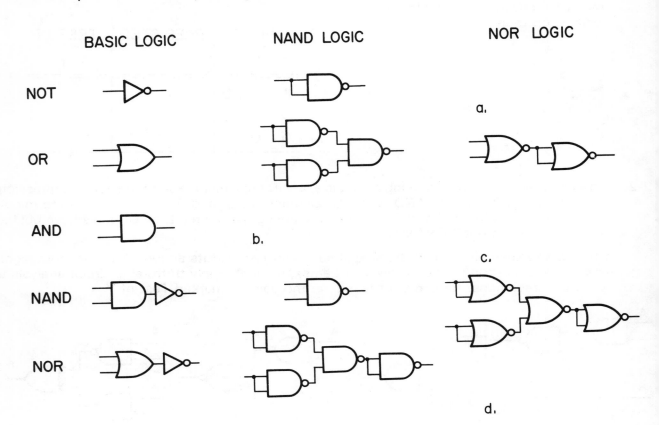

13-2. Using only NORs, draw the circuit represented by this expression. Do not simplify.

$$\overline{AB + C} = Y$$

13-3. Using only NANDs, draw the circuit represented by this expression. Do not simplify.

$$\overline{\overline{AB}} + C = Y$$

Chapter 5
NAND, NOR, XOR ELEMENTS

Name _____

Date _____ Score _____

Instructor _____

NAND AND NOR TRUTH TABLE CIRCUIT ANALYSIS

4-1. For the inputs shown, indicate the signals at the outputs of these logic elements.

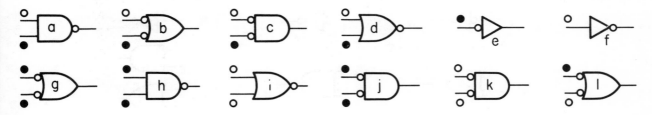

4-2. The output of this circuit is incorrect. Circle the lead where the error first appears.

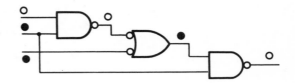

4-3. Repeat Problem 2 for this circuit and input signal set.

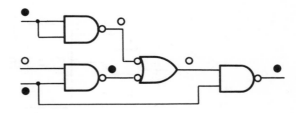

4-4. For the inputs shown, determine the output of this circuit.

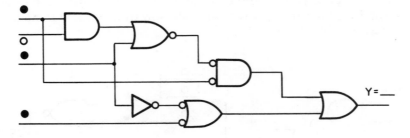

4-5. Repeat Problem 4 for this circuit and input signal set.

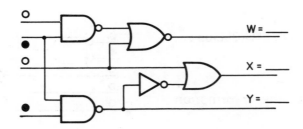

Chapter 5
NAND, NOR, XOR ELEMENTS

Name _____

Date _____ Score _____

Instructor _____

DE MORGAN'S THEOREMS

15-1. Match the following with the symbols at the right.

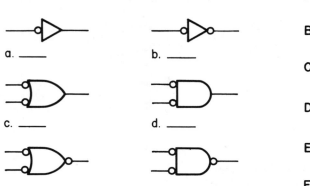

a. ____   b. ____

c. ____   d. ____

e. ____   f. ____

15-2. Which AND symbol gives the most direct indication of the states of A and B needed to activate ENABLE?

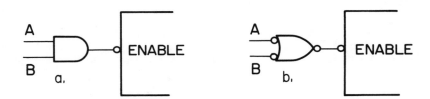

15-3. These are identical circuits. The symbols at the right are the DeMorgan implementation of those at the left. For the input signals shown (all 0s), determine the signals present at all points in both circuits. They should be identical.

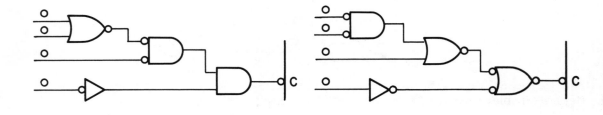

15-4. For the inputs shown, will ENABLE be active or inactive?
   a. Active (low).
   b. Inactive (high).
   c. Cannot tell from the information given.

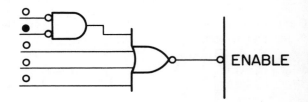

**Chapter 5
NAND, NOR, XOR ELEMENTS**

Name _____

Date _____ Score _____

Instructor _____

REVIEW

6-1. For the input signals shown, construct the output of this NAND.

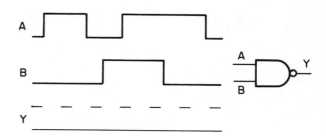

6-2. Draw a logic diagram to represent the following expression. Do not simplify. Use only three logic elements.

$$(\overline{A + B})(\overline{A + C}) = Y$$

W5-16-2

6-3. Write an expression to represent this circuit. Do not simplify.

_____ = Y

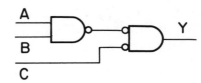

6-4. Which of the following expressions represents a circuit that contains an XOR?

a. $\overline{A}\overline{C} + A\overline{B} + B + \overline{A}B = Y$

b. $A\overline{B} + A(B + \overline{C}) = Y$

c. $A(\overline{B} + \overline{C}) + AC = Y$

W5-16-4

6-5. For the input signals shown, determine the output of this circuit.

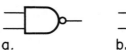

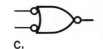

6-6. Which symbol at the right represents the DeMorgan implementation of an AND?

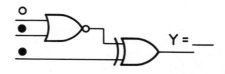

a.　　　　　　b.　　　　　　c.

6-7. This is the NOR equivalent of which of the following elements?
a. AND.
b. OR.
c. NAND.

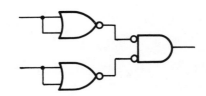

Chapter 6

DESIGN OF LOGIC CIRCUITS

Name _____

Date _____ Score _____

Instructor _____

SUBJECTIVE DESIGN (7404, 7408, 7432)

-1. Laboratory Activity. This circuit must be altered. A fourth input is to be added. When D = 0, the new output (Y^1) is to be 0. When D = 1, Y^1 is to equal Y (the present output). Add the necessary circuits to the logic diagram to accomplish this task.

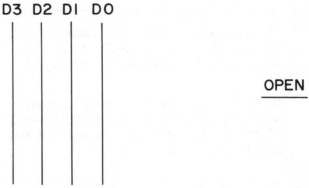

If this is an experiment, build your redesigned circuit and test its action in the laboratory.

-2. Laboratory Activity. When the number
D3D2D1D0 = 1101
appears on this bus, a 1 is to be output on lead OPEN. This output must be 0 for all other numbers on the bus. Draw the logic diagram for a decoder that will respond as indicated.

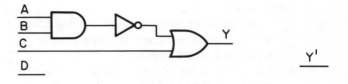

OPEN

If this is an experiment, build your circuit and test its action in the laboratory.

-3. Laboratory Activity. Three sensors are to drive the same alarm. If smoke is detected, the alarm is to sound. It is also to sound if both the temperature and relative humidity exceed their set points (90° and 70%) at the same time. The sensors and the alarm are active-high.

Design a circuit that will accomplish this control task, and draw its logic diagram in the space provided.

If this is an experiment, build your circuit and test its action in the laboratory.

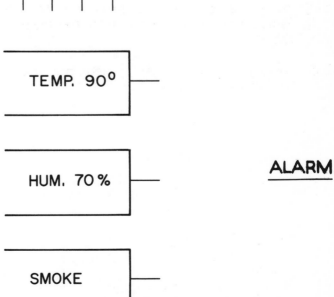

ALARM

Chapter 6
DESIGN OF LOGIC CIRCUITS

Name _____

Date _____ Score _____

Instructor _____

TRUTH TABLE FROM STATEMENT OF PROBLEM

2-1. Parts from three sources are delivered to this conveyor loading machine. Weights of parts from each source are:
A. 40 pounds.
B. 60 pounds.
C. 80 pounds.
This machine places parts on the conveyor according to the requirements of an automatic box loader at the right. Its requirements are:
a. Parts are not to be placed on the conveyor until the boxer signals that it is ready by setting R = 1.
b. Parts are to be delivered in groups with total weights of 100 pounds or more.
c. No more than one part from a given source is to be included in each group.
To meet requirement c, stops have been provided to allow only one part from each source to reach the loading points.

The presence of parts at the loading points are sensed by switches A, B, and C. When a part is present, these switches output 1s.

Complete the truth table to describe the relationship between A, B, C, R, and LOAD. A "1" on LOAD means that the loading shaft will rotate.

2-2. Write a Boolean expression that represents the requirements of the truth table.

2-3. Design a circuit that implements the Boolean expression.

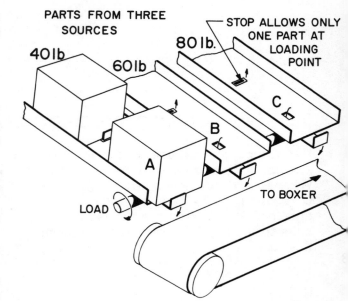

PARTS FROM THREE SOURCES

STOP ALLOWS ONLY ONE PART AT LOADING POINT

40 lb   60 lb   80 lb.

C

B

A

LOAD

TO BOXER

INPUTS				OUTPUT
40	60	80		
A	B	C	R	LOAD
0	0	0	0	
0	0	0	1	
0	0	1	0	
0	0	1	1	
0	1	0	0	
0	1	0	1	
0	1	1	0	
0	1	1	1	
1	0	0	0	
1	0	0	1	
1	0	1	0	
1	0	1	1	
1	1	0	0	
1	1	0	1	
1	1	1	0	
1	1	1	1	

Chapter 6
DESIGN OF LOGIC CIRCUITS

Name _____

Date _____ Score _____

Instructor _____

DESIGN OF A DIGITAL CIRCUIT (7404, 7408 or 4049, 4081)

BASE 2 INPUTS		BASE 10 OUTPUTS					
		PREDICTED			MEASURED		
AI	AO	Y3	Y2	YI	Y3	Y2	YI
O	O						
O	I						
I	O						
I	I						

BASE 2 INPUTS

AI —
AO—

BASE 10 OUTPUTS

— Y3
— Y2
— YI

8-1. Laboratory Activity. Based on the following description, complete the PREDICTED columns of the above truth table.

A decoder is to be designed that will accept a 2-bit binary number, A1AO, and output its decimal equivalent on leads Y3, Y2, and Y1. When A1 AO equals OO, Os are to appear on the three output leads. When A1 AO = O1, a 1 is to be output on lead Y1. When A1AO = 10, a 1 is to appear on lead Y2. When A1AO = 11, a 1 is to appear on lead Y3.

8-2. Laboratory Activity. Using sum-of-products, write expressions for the outputs of the above circuit. Do not simplify.

_____ = Y1   _____ = Y2   _____ = Y3

8-3. Laboratory Activity. Draw the logic diagrams represented by the above expressions. Do not simplify. Use ANDs and NOTs to implement the expressions.

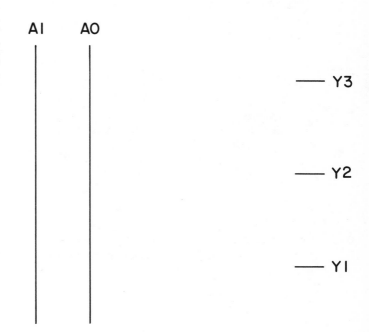

AI   AO

— Y3

— Y2

— YI

8-4. Laboratory Activity. Construct the circuit you have designed and test its action. Record your results in the columns marked MEASURED.

Chapter 6
DESIGN OF LOGIC CIRCUITS

Name _____

Date _____ Score _____

Instructor _____

DESIGN OF DIGITAL CIRCUITS (7404, 7408, 7432 or 4049, 4081, 4072

4-1. Laboratory Activity. Use sum-of-products to write the expression represented by this truth table. Do not simplify. Then implement your expression. If this is an experiment, construct the circuit and test your design. If this is not an experiment, use truth table circuit analysis to test the action of your circuit.

INPUTS		OUTPUT	TEST OF
A	B	Y	Y
O	O	O	
O	I	I	
I	O	O	
I	I	I	

4-2. Laboratory Activity. Repeat the experiment in Problem 1 using this truth table.

INPUTS			OUTPUT	TEST OF
A	B	C	Y	Y
I	O	I	I	
O	I	I	I	

OTHER INPUT SETS
RESULT IN Y = O.

4-3. Laboratory Activity. Repeat the experiment in Problem 1 using this truth table.

INPUTS			OUTPUT	TEST OF
A	B	C	Y	Y
I	O	I	I	
O	I	I	I	
I	O	O	I	

Y = O FOR OTHER INPUT SETS

Chapter 6
DESIGN OF LOGIC CIRCUITS

Name _____

Date _____ Score _____

Instructor _____

LAWS OF BOOLEAN ALGEBRA

6-1. Which of the following is *not* an acceptable application of the commutative law to the expression at the right?

$$AB + C = Y$$

    a. $BA + C = Y$

    b. $C + AB = Y$

    c. $BC + A = Y$

6-2. Which of the following is *not* an acceptable application of the associative law to the expression at the right?

$$\overline{A}\,\overline{BCD} = Y$$

    a. $(\overline{A}\,\overline{B})\overline{CD} = Y$

    b. $\overline{A}(\overline{BC})\overline{D} = Y$

    c. $\overline{A}\,\overline{B}(\overline{CD}) = Y$

6-3. Remove the brackets from each of the following expressions. Do not simplify.

    a. $A(B + C) =$ _____

    c. $(A + B)(C + D) =$ _____

    b. $AB(C + DE) =$ _____

    d. $A(BC + D)(E + F) =$ _____

6-4. Factor the largest possible number of inputs from each of the following. Do not attempt further simplification.

    a. $AB + AC + AD =$ _____

    c. $ABC + BC =$ _____

    b. $ABC + ACD =$ _____

    d. $ABC + ABD + DE =$ _____

6-5. The following are equivalent forms. That is, they have identical truth tables. In the space provided, implement each expression without additional simplification.

$$ABC + ABD = Y \qquad\qquad AB(C + D) = Y$$

Chapter 6
DESIGN OF LOGIC CIRCUITS

Name _____

Date _____ Score _____

Instructor _____

RULES OF TAUTOLOGY

The term *TAUTOLOGY* means a sort of repetition.

6-1. Complete the following identities. The first equation has been completed to serve as an example of the answers that will result.

    a. A + A = A
    b. AA = _____
    c. A + 1 = _____
    d. A1 = _____
    e. A + 0 = _____
    f. A0 = _____

6-2. Using the above identities, simplify the following expression. Show all steps, and implement your simplified expression.

$$A(A + B) = Y$$

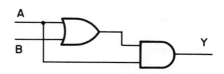

6-3. Repeat Problem 2 for this expression.

$$(AC + ABC)A = Y$$

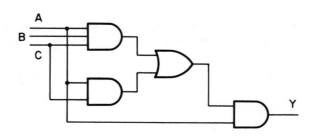

Chapter 6
DESIGN OF LOGIC CIRCUITS

Name _____

Date _____ Score _____

Instructor _____

SIMPLIFICATION (7408, 7432 or 4081, 4072)

-1. Laboratory Activity. Based on the following expression, complete the column in the truth table marked ORIGINAL. Then, simplify the expression and implement the new expression. If this is an experiment, build this circuit and experimentally test your simplification. If this is not an experiment, apply truth table circuit analysis to your diagram to test the results of the simplification. Record your results in the column marked SIMPLIFIED.

$AB + ABC + D = Y$

INPUTS				ORIGINAL	SIMPLIFIED
A	B	C	D	Y	Y
I	I	O	O		
O	I	I	O		
I	O	I	I		

-2. Laboratory Activity. Repeat the experiment in Problem 1 for this expression.

$AB(AB + BC)C = Y$

INPUTS			ORIGINAL	SIMPLIFIED
A	B	C	Y	Y
I	O	O		
I	I	I		
O	I	I		

Chapter 6
DESIGN OF LOGIC CIRCUITS

Name _____

Date _____ Score _____

Instructor _____

RELATED IDENTITIES FOR NOT ELEMENT

8-1. Complete the following identities.

a. $\bar{\bar{A}}$ = _____

b. $A\bar{A}$ = _____

c. $\bar{A} + A$ = _____

d. $\overline{A}\,\overline{B}$ = _____(DeMorgan)

e. $\bar{A} + \bar{B}$ = _____(DeMorgan)

8-2. Simplify this expression. Show all steps.

$$(\overline{A + B})(C + \bar{A}) = Y$$

8-3. Repeat Problem 2 for this expression.

$$(\overline{AB})(\overline{A + B}) = Y$$

8-4. Repeat Problem 2 for this expression.

$$(\overline{AB})(A + \bar{B})C = Y$$

**Chapter 6
DESIGN OF LOGIC CIRCUITS**

Name _____

Date _____ Score _____

Instructor _____

SIMPLIFICATION

9-1. Based on the following expressions, complete the two truth tables. Note that the expressions have identical truth tables. They are therefore equivalent.

$U + \bar{U}V = Y$ 　　　　　　　　$U + V = Y$

INPUTS		OUTPUT
U	V	Y
O	O	
O	I	
I	O	
I	I	

INPUTS		OUTPUT
U	V	Y
O	O	
O	I	
I	O	
I	I	

9-2. Complete these identities. They are used in the simplification of expressions.

a. $A + \bar{A}B =$ _____ 　　　　b. $\bar{A} + AB =$ _____

9-3. Simplify this expression. Show all steps.

$(AB + \bar{A}BC)B = Y$

9-4. Repeat Problem 3 for this expression.

$\bar{A}\bar{C} + A\bar{B}\bar{C} + C = Y$

Chapter 6
DESIGN OF LOGIC CIRCUITS

Name _____

Date _____ Score _____

Instructor _____

KARNAUGH MAPS

10-1. Based on this truth table, complete the following Karnaugh map.

	$\overline{C}\overline{D}$ 00	$\overline{C}D$ 01	CD 11	C$\overline{D}$ 10
$\overline{A}\overline{B}$ 00				
$\overline{A}B$ 01				
AB 11				
A$\overline{B}$ 10				

INPUTS				OUTPUT
A	B	C	D	Y
0	0	0	0	0
0	0	0	1	0
0	0	1	0	0
0	0	1	1	1
0	1	0	0	0
0	1	0	1	1
0	1	1	0	1
0	1	1	1	0
1	0	0	0	1
1	0	0	1	0
1	0	1	0	0
1	0	1	1	0
1	1	0	0	0
1	1	0	1	0
1	1	1	0	0
1	1	1	1	0

10-2. Based on the following Karnaugh map, complete this truth table.

	$\overline{C}$ 0	C 1
$\overline{A}\overline{B}$ 00	0	0
$\overline{A}B$ 01	1	1
AB 11	0	1
A$\overline{B}$ 10	1	0

INPUTS			OUTPUT
A	B	C	Y
0	0	0	
0	0	1	
0	1	0	
0	1	1	
1	0	0	
1	0	1	
1	1	0	
1	1	1	

Chapter 6
DESIGN OF LOGIC CIRCUITS

Name _____

Date _____ Score _____

Instructor _____

KARNAUGH MAPS

1-1. Write the expression and draw the logic diagram represented by this Karnaugh map. Try to get the simplest form.

_____ = Y

		$\overline{C}\overline{D}$ 00	$\overline{C}D$ 01	CD 11	$C\overline{D}$ 10
$\overline{A}\overline{B}$	00	0	0	0	0
$\overline{A}B$	01	0	0	1	0
AB	11	0	0	0	0
$A\overline{B}$	10	1	0	0	0

1-2. Repeat Problem 1 for this map.

_____ = Y

		$\overline{C}$ 0	C 1
$\overline{A}\overline{B}$	00	0	1
$\overline{A}B$	01	0	1
AB	11	0	0
$A\overline{B}$	10	0	0

1-3. Repeat Problem 1 for this map. Blanks imply 0s.

_____ = Y

		$\overline{C}\overline{D}$ 00	$\overline{C}D$ 01	CD 11	$C\overline{D}$ 10
$\overline{A}\overline{B}$	00				
$\overline{A}B$	01	1	1		
AB	11		1		
$A\overline{B}$	10			1	1

1-4. Repeat Problem 1 for this map.

_____ = Y

		$\overline{C}\overline{D}$ 00	$\overline{C}D$ 01	CD 11	$C\overline{D}$ 10
$\overline{A}\overline{B}$	00		1		
$\overline{A}B$	01				
AB	11				
$A\overline{B}$	10		1	1	

**Chapter 6
DESIGN OF LOGIC CIRCUITS**

Name _____

Date _____ Score _____

Instructor _____

KARNAUGH MAPS

12-1. Write the expression and draw the logic diagram represented by this Karnaugh map. Try to get the simplest form. Zeros have been omitted to emphasize the 1s.

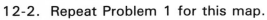 = Y

		C̄D̄ 00	C̄D 01	CD 11	CD̄ 10
ĀB̄	00	I	I		
ĀB	01	I	I		I
AB	11				
AB̄	10				

12-2. Repeat Problem 1 for this map.

_____ = Y

		C̄D̄ 00	C̄D 01	CD 11	CD̄ 10
ĀB̄	00		I	I	
ĀB	01				
AB	11				
AB̄	10	I	I	I	

12-3. Repeat Problem 1 for this map. Ds stand for "don't cares." Substitute 1s and 0s for the Ds to produce the simplest expression.

_____ = Y

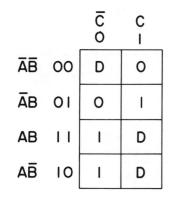

		C̄ 0	C 1
ĀB̄	00	D	0
ĀB	01	0	I
AB	11	I	D
AB̄	10	I	D

12-4. Repeat Problem 1 for this map.

_____ = Y

		C̄D̄ 00	C̄D 01	CD 11	CD̄ 10
ĀB̄	00	I	I	I	I
ĀB	01	D	I	I	D
AB	11	0	I	0	0
AB̄	10	0	D	0	0

Name _____

Date _____ Score _____

Instructor _____

DIGITAL CIRCUIT DESIGN (7404, 7408, 4732 or 4049, 4081, 4072)

3-1. Use sum-of-products to write an expression representing this truth table. Then simplify it. Show all steps.

INPUTS			OUTPUT	TEST OF
A	B	C	Y	Y
0	0	0	I	
0	0	I	I	
0	I	0	I	
0	I	I	I	
I	0	0	O	
I	0	I	I	
I	I	0	O	
I	I	I	O	

13-2. Laboratory Activity. Complete this Karnaugh map using data from the above truth table. Then use the map to write the expression it represents. Compare this expression with the one obtained by using sum-of-products and simplification. Implement this expression. If this is an experiment, build the circuit and test its action. Record your results in the column marked TEST OF Y.

_____ = Y

	$\overline{C}$ 0	C I
$\overline{A}\overline{B}$ 00		
$\overline{A}B$ 01		
AB II		
A$\overline{B}$ 10		

3-3. Laboratory Activity. Based on this Karnaugh map, write the simplest expression possible. Then draw the circuit it represents. If this is an experiment, build the circuit and test its action. Record your results in the second map.

_____ = Y

	$\overline{C}$ 0	C I
$\overline{A}\overline{B}$ 00	D	O
$\overline{A}B$ 01	O	I
AB II	I	D
A$\overline{B}$ 10	O	O

	$\overline{C}$ 0	C I

**Chapter 6
DESIGN OF LOGIC CIRCUITS**

Name _____

Date _____ Score _____

Instructor _____

REVIEW (7400, 7402, 7404, 7408, 7432, 7486)

14-1. Laboratory Activity. Write and simplify the expression for this circuit. Then draw the circuit represented by the simplified expression. If this is an experiment, construct both circuits and determine if their outputs are identical. If this is not an experiment, use truth table circuit analysis to test the original and simplified circuits. Record your results in the proper columns.

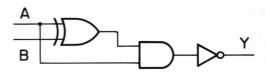

INPUTS		ORIGINAL	SIMPLIFIED
A	B	Y	Y
O	O		
O	I		
I	O		
I	I		

14-2. Laboratory Activity. Complete this Karnaugh map using the data in this truth table. Ds stands for don't cares. Use the map to write the simplest possible expression. Then draw the circuit represented by the expression. If this is an experiment, build the circuit and test its action. If this is not an experiment, use truth table circuit analysis to test the circuit. Record your results in the column marked TEST OF Y.

_____ = Y

INPUTS			OUTPUT	TEST OF
A	B	C	Y	Y
O	O	O	D	
O	O	I	O	
O	I	O	I	
O	I	I	D	
I	O	O	I	
I	O	I	O	
I	I	O	I	
I	I	I	D	

		$\bar{C}$ O	C I
$\bar{A}\bar{B}$	OO		
$\bar{A}B$	OI		
AB	II		
A$\bar{B}$	IO		

Chapter 7

SEQUENTIAL LOGIC

Name _____

Date _____ Score _____

Instructor _____

RELAY-BASED MEMORY ELEMENTS

-1. Use a series of arrows to show the path (a complete circuit) of the current through the coil of the relay. Do this for a time after 2PB has been pressed and released.

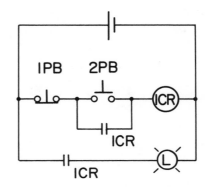

-2. Repeat Problem 1 for the loop containing the lamp.

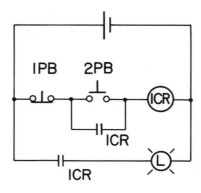

-3. Complete this truth table for the above circuit. A 1 in an input column implies the pressing and releasing of a switch. A 1 in the output column means that the lamp is lit. A star (*) indicates an undefined condition. The label Lt implies a previous value.

INPUTS		OUTPUT
1PB	2PB	Lt+1
0	1	
1	0	
0	0	
1	1	

1-4. This circuit is used to start and stop a small ac motor. The symbol M represents a contactor (a large relay with contacts capable of carrying the motor current). Circle the set of contacts that would be called holding contacts.

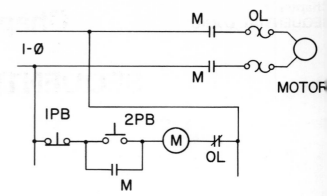

1-5. Refer to the diagram used in Problem 4. Place an X on the symbol representing the coil of the contactor.

Chapter 7
SEQUENTIAL LOGIC

Name _____

Date _____ Score _____

Instructor _____

FLIP-FLOP

7-1. Complete this truth table for an R-S flip-flop. Entries are not in standard order. Use a star (*) to indicate undefined outputs. Use Qt to indicate a previous value of Q.

INPUTS		OUTPUT
S	R	Qt+1
I	I	
O	I	
O	O	
I	O	

7-2. For the values of R, S, and Q shown at the right, on which flip-flop input did the last 1 appear?
a. On S.
b. On R.
c. Cannot tell from the information given.

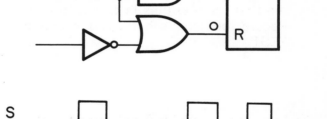

7-3. For the inputs shown, carefully sketch the expected output of an R-S flip-flop such as the one at the top of this sheet. Assume Q = 0 at t = 0.

S

R

Q

7-4. Sketch the simplest set of input signals that would result in the indicated output. Assume S, R, and Q are all 0 at t = 0.

S

R

Q

7-5. Which of the following best describes Q after time t3?
a. Will follow path A (Q = 1).
b. Will follow path B (Q = 0).
c. Could follow either path A or B.

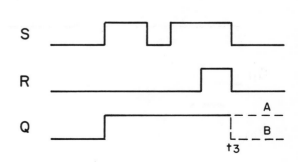

S

R

Q

101

Chapter 7
SEQUENTIAL LOGIC

Name _____

Date _____ Score _____

Instructor _____

NAND-BASED FLIP-FLOP (7400 or 4011)

3-1. Laboratory Activity. Complete the column of this truth table marked PREDICTED.

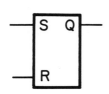

INPUTS		PREDICTED	TEST OF
S	R	Qt+1	Qt+1
I	O		
O	I		
O	O		
I	I		

3-2. Laboratory Activity. Construct this NAND-based flip-flop and experimentally verify the first three entries in the truth table. Record your results in the column marked TEST.

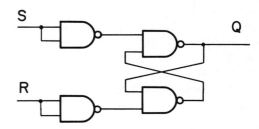

3-3. Laboratory Activity. To observe the effect of SET and RESET being active at the same time (the last row in the truth table), place 1s on both S and R. Then manually switch them to 0 at the same time. Repeat this effort a number of times and record the results in the following table. Discuss the results with your partner.

TRIAL	I	2	3	4	5	6	7	8
Qt+1								

3-4. Laboratory Activity. Predict the output of the above R-S flip-flop for the following inputs. Record your predictions on graph Qp. Then apply these input signals to the flip-flop you have constructed and verify your predictions. Record your results on the graph marked Qm.

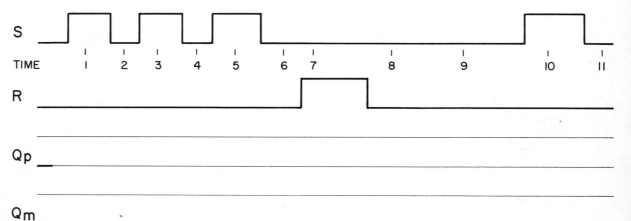

102

Chapter 7
SEQUENTIAL LOGIC

Name _____

Date _____ Score _____

Instructor _____

NOR-BASED FLIP-FLOP (7402 or 4001)

7-1. Laboratory Activity. Based on the expected action of an R-S flip-flop, complete the column of this truth table labeled PREDICTED. Then build the following circuit and experimentally show that it is an R-S flip-flop. Record your results in the column marked TEST OF Q_{t+1}.

INPUTS		PREDICTED	TEST OF
S	R	Q_{t+1}	Q_{t+1}
I	O		
O	I		
O	O		
I	I		

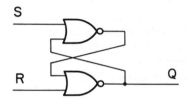

7-2. Laboratory Activity. Input signals and some values of Q are shown on the flip-flops at the right. Use truth table circuit analysis to determine the expected values of A and Q. Record your predictions in the blanks marked Ap and Qp.

Then, apply the indicated signals to your circuit and verify your predictions. Record the results in the blanks marked Am and Qm.

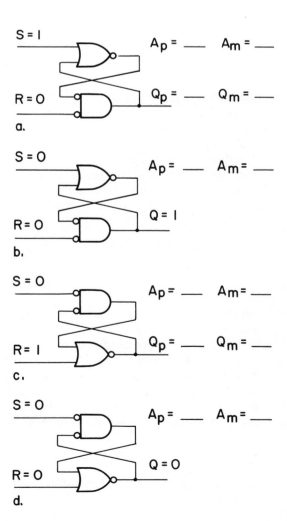

**Chapter 7
SEQUENTIAL LOGIC**

Name _____

Date _____ Score _____

Instructor _____

FLIP-FLOP WITH ACTIVE-LOW INPUT (7400 or 4011)

5-1. Complete the logic diagram of a NAND-based flip-flop. It is to have active-low inputs.

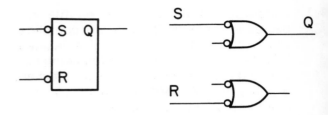

5-2. Laboratory Activity. Predict the action of the above circuit and record those predictions in the appropriate column of this truth table. If this is an experiment, build the circuit and experimentally test your predictions. Record your results in the column marked TEST OF Q_{t+1}.

INPUTS		PREDICTED	TEST OF
S	R	Q_{t+1}	Q_{t+1}
0	1		
1	0		
1	1		
0	0		

5-3. Laboratory Activity. For the inputs shown, draw the expected output of the above flip-flop. Assume $Q = 0$ at $t = 0$. Use graph Q_p. If this is an experiment, apply the indicated signals to the circuit you built and test your predictions. Record your results on the graph marked Q_m.

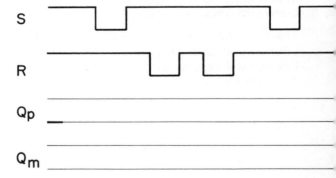

5-4. Laboratory Activity. Assume that the active-low inputs of this flip-flop are floating. If the indicated ground lead was momentarily touched to R, what signal would appear at Q? If this is an experiment, disconnect the leads to R and S of the circuit you have built. Then, alternately touch a grounded lead to R and S to test your prediction.

a. Q will go to 1.
b. Q will go to 0.
c. There is no way of telling which number (1 or 0) will appear at Q.

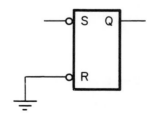

Chapter 7
SEQUENTIAL LOGIC

Name _____

Date _____ Score _____

Instructor _____

APPLICATION OF FLIP-FLOPS

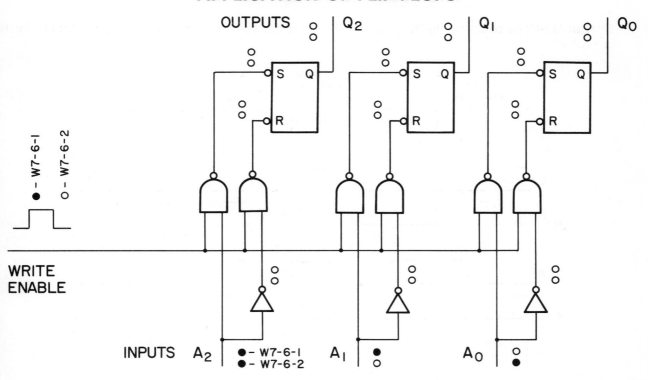

A 3-bit register is shown. Data enters through A2A1A0. Stored data is available at Q2Q1Q0. To enter a new number, WRITE ENABLE must go active (high). When WRITE ENABLE is low, the number at A cannot enter.

6-1. Use the upper dot in each pair. Upper dots are marked W7-6-1. Lower dots are marked W7-6-2. For the input shown (upper dots) and WRITE ENABLE active, indicate the numbers present at the various points in the circuit. Do this by filling in the dots where 1s will appear.

6-2. Repeat Problem 1 (related to ''Problem 7-6-1'' and ''Problem 7-6-2'') for WRITE ENABLE inactive. Use the lower set of dots. Note that a new number appears at input A.

6-3. Change the above circuit into a 4-bit register. The input to the added bit is to be called A3. Its output is Q3.

6-4. Which of the following circuits would most likely be used for the flip-flops in the above register?

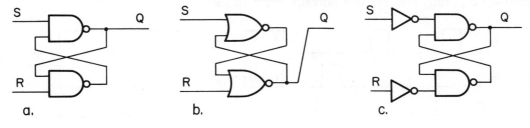

Chapter 7
SEQUENTIAL LOGIC

Name _____

Date _____ Score _____

Instructor _____

TROUBLESHOOTING FLIP-FLOPS

7-1. Output Q of one of these flip-flops is s-a-0. The other two are working properly. Which flip-flop is faulty?

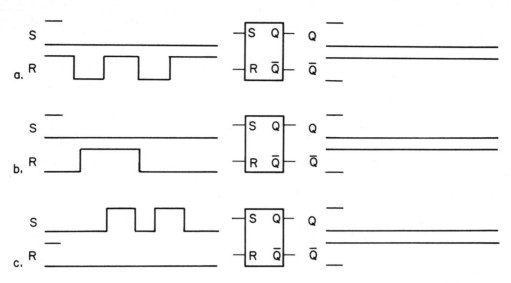

7-2. At the output of these flip-flops, Q and $\bar{Q}$ are both 1 at time t1. In one case the fault is in the driving circuits; in the other, the flip-flop is faulty. Which flip-flop is faulty?

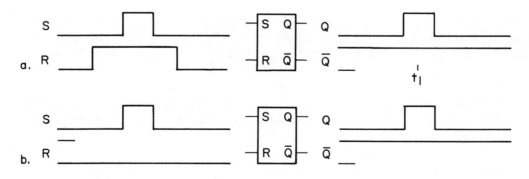

7-3. The output of this flip-flop is incorrect. The error was caused by a glitch. Draw this glitch in the correct position and on the correct input graph.

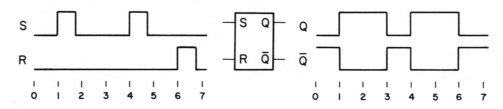

Name _____

Date _____ Score _____

Instructor _____

REVIEW OF R-S FLIP-FLOPS

8-1. Which diagram represents a sequential-logic circuit?

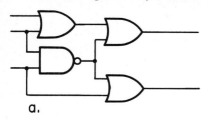

a.

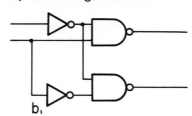

b.

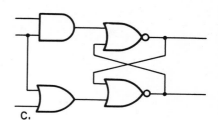

c.

8-2. Why are the output leads of this flip-flop
crossed at A?
a. To provide the necessary feedback.
b. To match the positions of the outputs
to those on the symbol.
c. To insure that both output leads are
the same length.

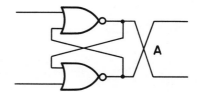

8-3. If bubble notation were used at the output of this flip-flop, on which lead would
the bubble be placed?
a. Q
b. $\overline{Q}$
c. On both Q and $\overline{Q}$.

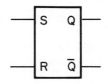

8-4. Draw the NAND-based circuit represented
by this R-S flip-flop symbol.

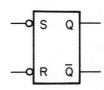

8-5. For the inputs shown, draw the output of this flip-flop. Assume $Q = 0$ at $t = 0$.

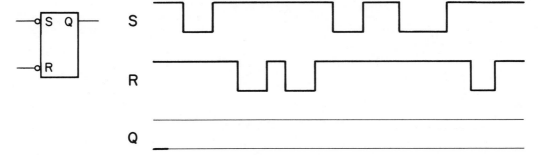

Chapter 8
CLOCKED FLIP-FLOPS

Name _____

Date _____ Score _____

Instructor _____

CLOCKED R-S FLIP-FLOPS

8-1. Complete this timing diagram for the flip-flop at the right.

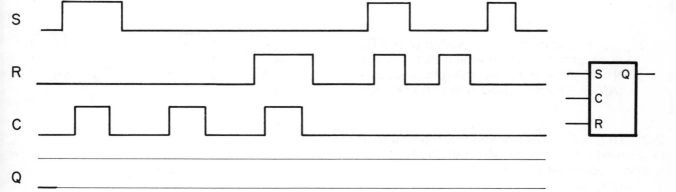

8-2. Repeat Problem 1 for this timing diagram and flip-flop. Note that this flip-flop has an active-low clock.

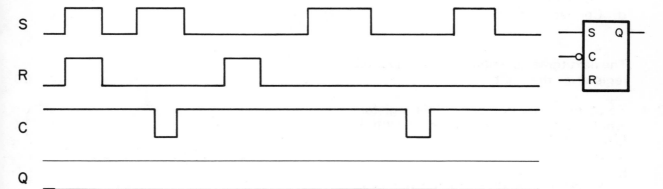

8-3. Repeat Problem 1 for this timing diagram. Note that the active-high, level-triggered clock is active for a long period of time.

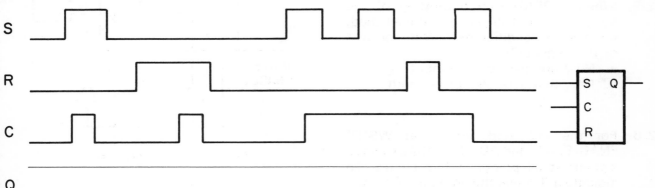

Chapter 8
CLOCKED FLIP-FLOPS

Name _____

Date _____ Score _____

Instructor _____

CLOCKED R-S FLIP-FLOPS

2-1. Which symbol represents the NAND-based flip-flop at the right?

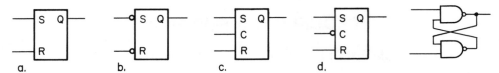

2-2. Clock circuits are to be added to the flip-flops at the right. In which circuit would NANDs be used in the clock circuit? (ANDs would be used in the other circuit.)

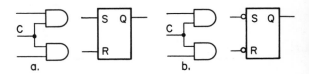

2-3. To convert this clocked R-S flip-flop to an unclocked R-S flip-flop, which of the following actions would be taken?
   a. Ground C.
   b. Let C float or connect it to a permanent logic 1.
   c. Connect C to S.
   d. Connect C to R.

The next three questions refer to the 2-bit register at the right.

2-4. To store a new number in this register, what signal must be applied to WRITE ENABLE?
   a. 0.
   b. 1.
   c. Either a 1 or 0, since the circuit is always in its write mode.

2-5. When a READ (the outputting of the stored number) is being accomplished, what is the condition of the input portion of this register?
   a. New numbers can be input.
   b. New numbers cannot be input.

2-6. For the indicated signals at WRITE ENABLE, A1, and A0, show the expected signals at all points in this circuit. To indicate a 0, leave the dot open; fill in the dot if a 1 is expected.

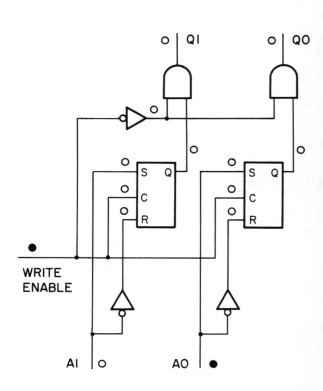

110

**Chapter 8
CLOCKED FLIP-FLOPS**

Name _____

Date _____ Score _____

Instructor _____

CLOCKED R-S FLIP-FLOPS (7400 or 4001)

8-1. Laboratory Activity. In the space at the right, draw the circuit for a NAND-based clocked R-S flip-flop. Label its leads S, R, C, and Q.

8-2. Laboratory Activity. Using graph Qp, predict the output of the above flip-flop. Then construct the circuit and experimentally test your predictions. Record your results on Qm.

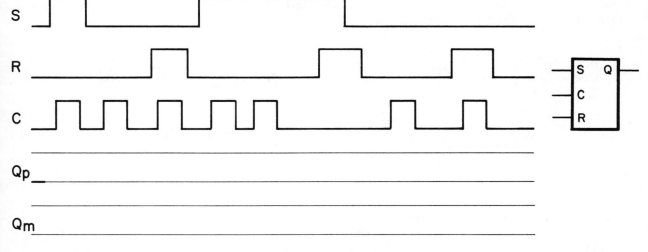

8-3. Laboratory Activity. Repeat Problem 2 for this set of inputs.

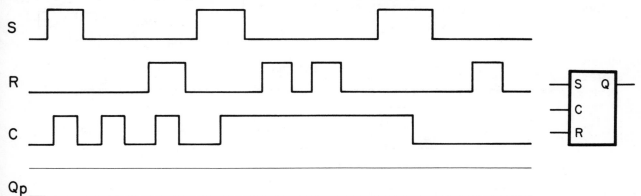

Chapter 8
CLOCKED FLIP-FLOPS

Name _____

Date _____ Score _____

Instructor _____

CLOCKED R-S FLIP-FLOPS (2-7400, 7402 or 2-4011, 4001)

4-1. Laboratory Activity. Which symbol best describes the action of the NOR-based flip-flop? If th
is an experiment, build this circuit and test your selection.

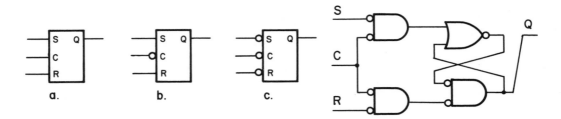

4-2. Laboratory Activity. For the input signals shown, indicate the expected signals at all points
this NAND-based flip-flop. Use the upper dot in each pair. If a 0 is expected, leave the dot ope▮
fill it in if a 1 is expected. If this is an experiment, build the circuit and test your prediction
Use the lower dots to record your results.

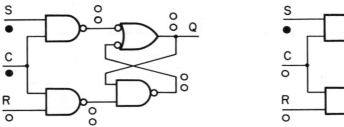

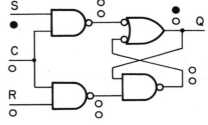

4-3. Laboratory Activity. This NAND-based
flip-flop has an active-low clock. Predict
its output for the following inputs. Record
your prediction on graph Qp. If this is an
experiment, build this circuit and test your
predictions. Use graph Qm.

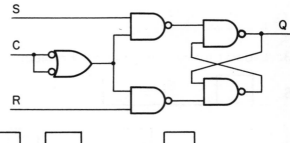

Chapter 8
CLOCKED FLIP-FLOPS

Name_____

Date _____ Score _____

Instructor _____

DATA LATCHES (2-7400 or 2-4011)

-1. Laboratory Activity. In the space at the right, draw the circuit for a data latch. Start with a NAND-based R-S flip-flop. Then use a NAND (as a NOT) to convert your circuit into a data latch.

-2. Laboratory Activity. Using graph Qp, predict the output of the above fip-flop for the following input signals. If this is an experiment, build the circuit and experimentally test your predictions. Use graph Qm to record your results.

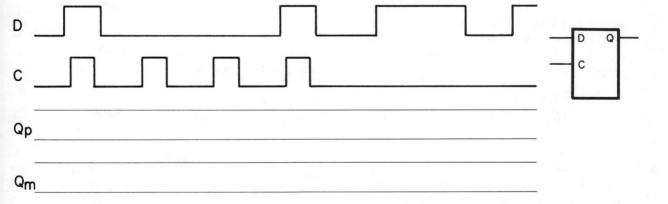

-3. Laboratory Activity. Repeat Problem 2 for this set of input signals.

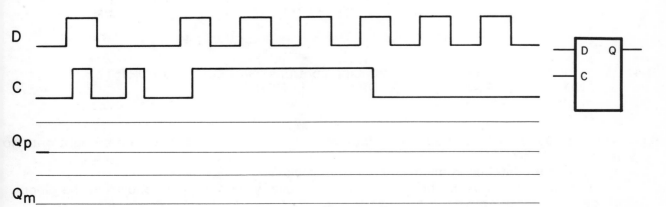

Chapter 8
CLOCKED FLIP-FLOPS

Name _____

Date _____ Score _____

Instructor _____

DATA LATCH APPLICATIONS

The circuit at the right is an output from a microprocessor-based computer. The pinout for the 7475 is shown below. When a number is to be output, it is placed on the data bus. Control leads OUT and PORT 1 are then made active. This latches the number in the 7475s and causes the number to appear at the circuit's output, DO7 through DO0.

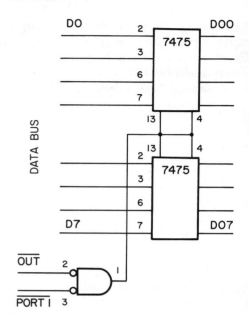

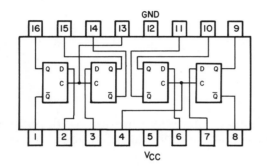

6-1. Why are two 7475s used in this circuit?
 a. One acts as a backup for the other.
 b. The computer's word size is 8 bits, so eight data latches are needed.
 c. Used in pairs, data latches can output both positive and negative logic levels.

6-2. Complete the above diagram by adding pin numbers to the output leads of the data latches

6-3. This is the timing diagram for data bus lead DO and the two control leads. At the end of the period shown, what number will be at DO0?
 a. 1.
 b. 0.

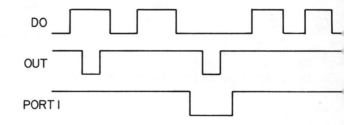

6-4. What logic element was used to combine control signals OUT and PORT 1?
 a. AND. b. NAND. c. OR. d. NOR.

6-5. Although each 7475 contains four data latches, only two clock leads are brought out. How is this possible?
 a. Two of the latches in each 7475 are unclocked.
 b. All four clock leads are brought out. For simplicity, only two are shown in the pinout.
 c. Each clock lead drives two data latches.

114

Chapter 8
CLOCKED FLIP-FLOPS

Name_____

Date _____ Score _____

Instructor _____

DATA LATCH APPLICATIONS

Problems on this sheet refer to the small scratch-pad memory shown.

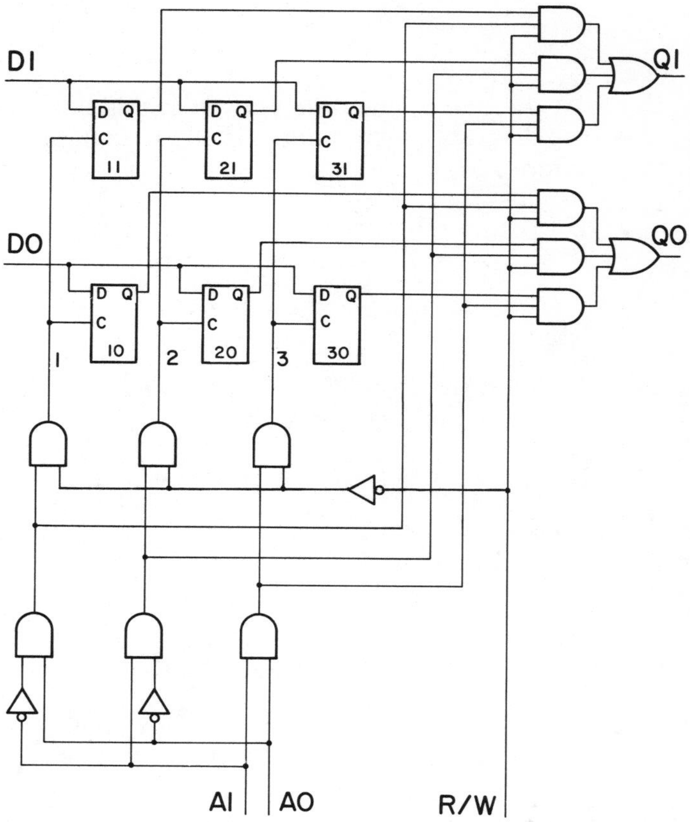

-1. What is the word size of this memory? That is, how many bits are there in each stored word?
   a. 1.
   b. 2.
   c. 3.
   d. 4.

-2. How many words can be stored in this memory?
   a. 2.
   b. 3.
   c. 4.
   d. 5.

-3. Which leads are used to input data?
   a. D1, D0.
   b. A1, A0.
   c. R/W.
   d. Q1, Q0.

-4. To store a word in flip-flops 11 and 10, what address must be applied to A1 and A0?
   a. 00.
   b. 01.
   c. 10.
   d. 11.

-5. To store a number in this memory, what signal must be placed on the control lead R/W?
   a. 1.
   b. 0.

115

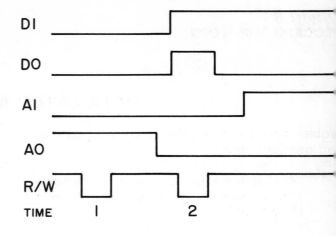

7-6. At times 1 and 2 in this timing diagram, numbers are stored. (Refer to R/W.) At which time is the number stored in data latches 11 and 10?
  a. Time 1.
  b. Time 2.
  c. At neither time 1 nor time 2.

7-7. At the end of the time shown in this timing diagram, what number is stored in data latch 10?
  a. 1.
  b. 0.

7-8. Including Vcc and GND, how many leads enter this memory circuit?
  a. 7.
  b. 8.
  c. 9.
  d. 10.

Name_____

Date _____ Score _____

Instructor _____

DATA LATCH APPLICATIONS

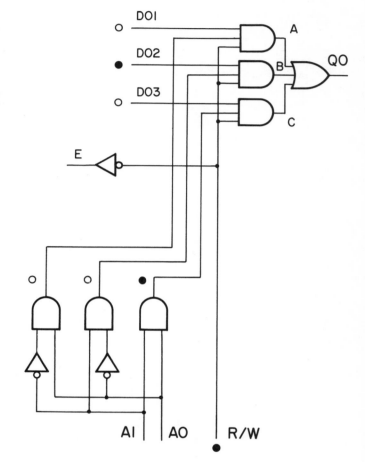

Shown is a portion of the output circuit from the memory on Sheet 8-7 (Chapter 8, numbers 7-1 through 7-8). Problems on this sheet refer to this drawing.

-1. For the signal shown at the inputs of the ANDs at the upper right, what numbers will appear at points A, B, C, and QO?

a. A = _____.

b. B = _____.

c. C = _____.

d. QO = _____.

-2. For the signals of Problem 1, which stored number appears at output QO?
a. DO1.
b. DO2.
c. DO3.
d. None of these.

-3. For the signals of Problem 1, what address must be applied to A1A0?
a. 00.
b. 01.
c. 10.
d. 11.

-4. Which group of ANDs is most likely referred to as a DECODER?
a. Those at the upper right.
b. Those at the lower left.

-5. When a number is being written (R/W = 0), what number appears at output QO?
a. Always 1.
b. Always 0.
c. The number that was output during the last read (output).
d. The new number being written (stored) in the memory.

-6. If A1A0 = 10, which stored number will appear at QO?
a. DO1.   b. DO2.   C. DO3.   d. None of these.

-7. Which of the following best describes the function of lead E?
a. Read enable.
b. Write enable.
c. Address enable.

Chapter 9

MASTER-SLAVE FLIP-FLOPS

Name _____

Date _____ Score _____

Instructor _____

MASTER-SLAVE FLIP-FLOP TIMING DIAGRAMS

-1. Use graph Q0 to show the expected output of this ordinary data latch. That is, it is not master-slave. Then use graph Qms to show the expected output of a master-slave data latch. Assume both are level triggered with active-low clocks. At time t = 0, Q = 0.

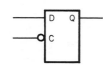

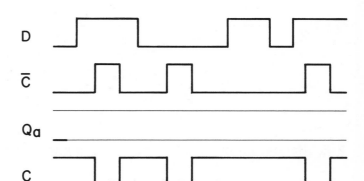

-2. Complete this timing diagram for this master-slave flip-flop. It has been constructed using two ordinary flip-flops. Clock Ca equals $\overline{C}$, while Cb equals C. Assume Q = 0 at t = 0.

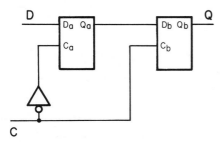

-3. Complete this timing diagram for the following master-slave level-triggered flip-flop. Note that the clock is active for a long period near the middle of the diagram. Assume Q = 0 at t = 0.

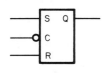

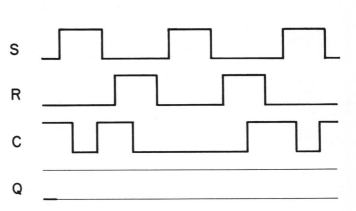

Chapter 9
MASTER-SLAVE FLIP-FLOPS

Name _____

Date _____ Score _____

Instructor _____

SHIFT REGISTER

2-1. Complete the timing diagram for the shift register shown. The data latches are active-low an level-triggered. Assume all outputs are O at t = O.

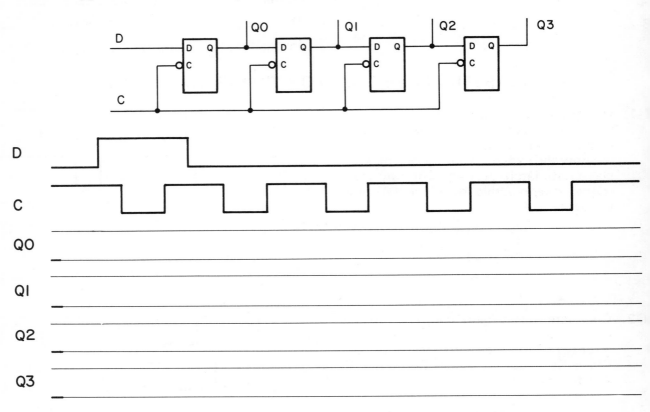

2-2. Repeat Problem 1 for the set of inputs shown.

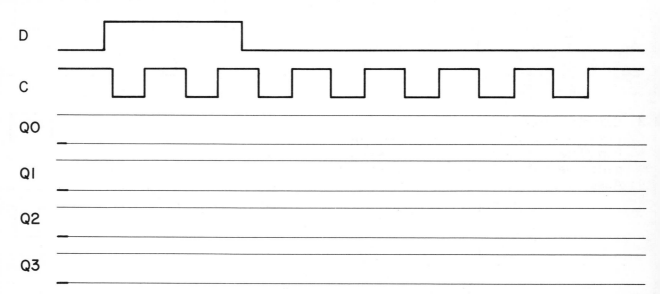

Chapter 9
MASTER-SLAVE FLIP-FLOPS

Name _____

Date _____ Score _____

Instructor _____

R-S MASTER-SLAVE FLIP-FLOP (2-7400, 7404 or 2-4011, 4049)

9-1. Laboratory Activity. Use graphs Ap and Qp to indicate your predictions of the signals at A and Q in this circuit. Assume A and Q are at 0 when t = 0.

9-2. Laboratory Activity. Build this circuit. Be sure to indicate pin numbers on the logic diagram. Also, build and test the first flip-flop before adding the second. This will make troubleshooting easier. Use a debounced signal source (called a pulser on some trainers) for C. Record your results on graphs Am and Qm.

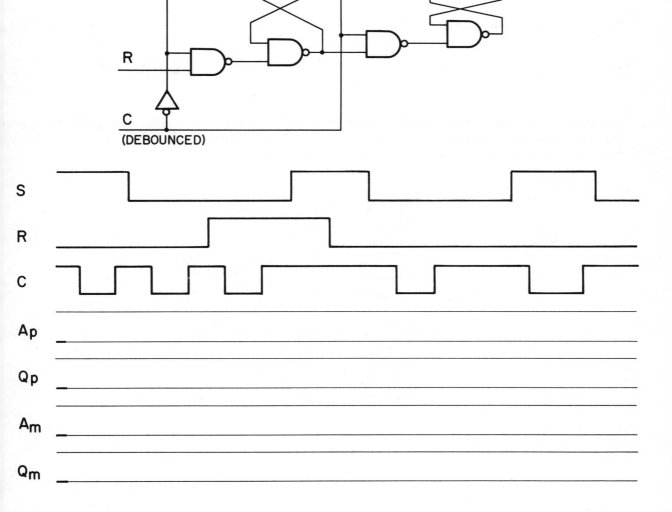

Chapter 9
MASTER-SLAVE FLIP-FLOPS

Name _____

Date _____ Score _____

Instructor _____

EDGE-TRIGGERED DATA LATCH (7410, 7400 or 4023, 4011)

4-1. Laboratory Activity. For the inputs shown, predict the output of this edge-triggered data latch. Its clock is active on its positive-going edge. Record your predictions on graph Qp.

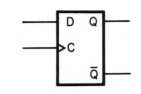

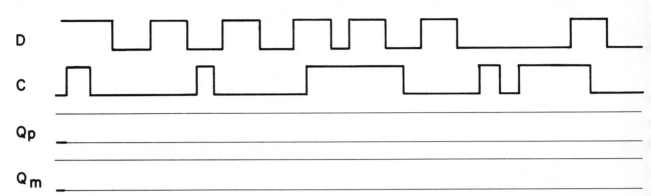

4-2. Laboratory Activity. Build the following data latch and test your predictions. Circuits as complex as this should be built and tested in sections. To aid in troubleshooting, pin numbers should be noted on the logic diagram. Record your results on graph Qm.

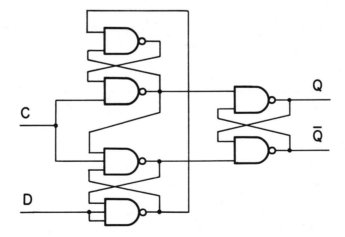

Chapter 9
MASTER-SLAVE FLIP-FLOPS

Name _____

Date _____ Score _____

Instructor _____

SHIFT REGISTER (2-7474 or 2-4013)

5-1. Laboratory Activity. For the inputs shown, predict the output of this shift register. Use graphs Q_{0P} through Q_{3P}. The data latches have positive-going edge-triggered clocks. Assume all outputs are 0 at t = 0.

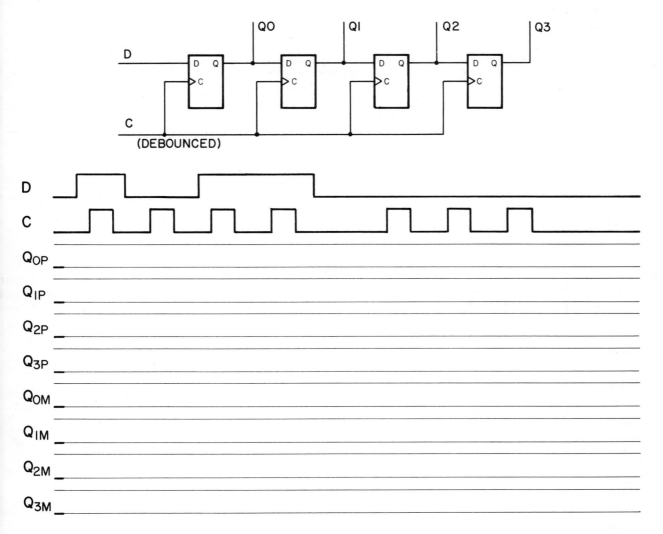

5-2. Laboratory Activity. Build the shift register shown and experimentally test your predictions. Record your results on the remaining graphs.

Construction Notes: The 7474 and 4013 have preset and clear inputs. On the 7474, these inputs are usually marked PR and CLR. On the 4013 they may be marked SET and RESET. On the 7474, PR and CLR are active-low. Because floating TTL leads act like logic 1s, these inputs can often be left unconnected in the laboratory. If erratic operation results, however, these leads must be connected to +5 V. If the 4013 is used, SET and RESET are active-high. They must be grounded to prevent improper operation.

Chapter 9
MASTER-SLAVE FLIP-FLOPS

Name _____

Date _____ Score _____

Instructor _____

RING COUNTER (2-7474, 7427 or 7402 or use 2-4013, 4025 or 4011)

6-1. Laboratory Activity. Predict the output of this ring counter. Use graphs Q0P through Q3P. Not that input D is not complete on the graph. Your predictions should include the signals on thi lead. Assume all outputs are as shown at t = 0.

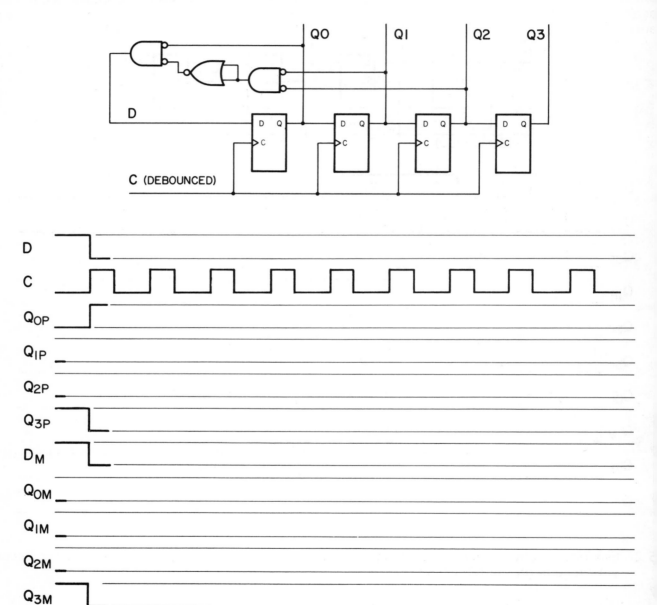

6-2. Laboratory Activity. Build this ring counter and test your predictions. The circuit from Probler 5-2 can be modified to produce this circuit. The three NORs simulate a 3-input element. If th 3-input NOR is available it can be used to simplify this circuit. Record your results on the remainin graphs.

Name _____

Date _____ Score _____

Instructor _____

SHIFT REGISTER APPLICATION

Shown is a register from a small computer. It has parallel inputs and outputs. In addition to temporarily storing data, it can SHR (shift right) and ROR (rotate right). When SHR is accomplished, stored bits are shifted to the right one place. Zeros are entered at the left. When ROR is accomplished, bits are again shifted one place to the right. However, the LSB is brought from the right of the register and entered at the left. That is, the LSB replaces the MSB.

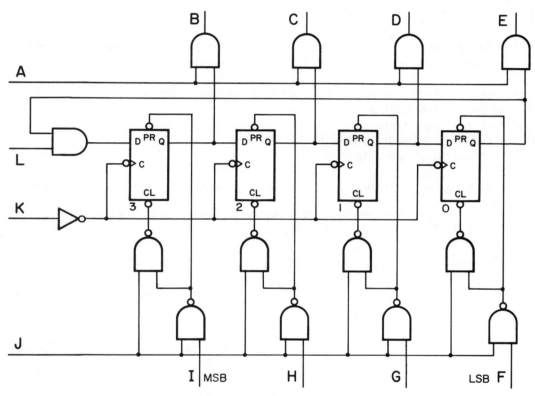

7-1. _____ Which lead is INPUT ENABLE? That is, which lead must be active to store a new number?

7-2. INPUT ENABLE is:
   a. Active-high.
   b. Active-low.

7-3. _____ Which lead is OUTPUT ENABLE?

7-4. OUTPUT ENABLE is:
   a. Active-high.
   b. Active-low.

7-5. When OUTPUT ENABLE is inactive, what numbers appear at the circuit's output?
   a. All 0s.
   b. All 1s.
   c. The last number stored.

7-6. T   F   To store a new number, the clocks on the data latches must be activated by making input K high. Circle the correct answer.

Chapter 9
MASTER-SLAVE FLIP-FLOPS

Name _____

Date _____ Score _____

Instructor _____

SHIFT REGISTER APPLICATIONS

Problems on this sheet refer to the circuit on the sheet for Problem 9-7 (7-1 through 7-6).

8-1. _____ Which lead determines whether a shift or rotate will be accomplished?

8-2. Which designation would probably be used on the lead described in Problem 8-1?
   a. SHIFT/ROTATE.
   b. ROTATE/SHIFT.

8-3. T   F   To accomplish a shift or rotate, the clocks of the data latches must be activated b
            making lead K high. Circle the correct answer.

8-4. Which best describes the action of the clocks on the data latches?
   a. Active-high, level-triggered.
   b. Active-low, level-triggered.
   c. Positive-going, edge-triggered.
   d. Negative-going, edge-triggered.

8-5. T   F   The flip-flops in this circuit must be master-slave. Circle the correct answer.

8-6. T   F   With only minor changes, R-S flip-flops could be substituted for data latches 0, 1
            and 2. Circle the correct answer.

8-7. Which NOT symbol is preferred in lead K?

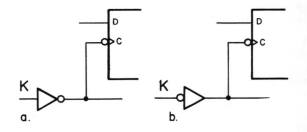

8-8. This register can be cleared (outputs set to 0) by activating lead K four times. To use this ap
     proach to clearing this register, what signal must be on the lead that selects shift or rotate
   a. SHIFT.
   b. ROTATE.

8-9. At the right, a NAND has been used in
     place of the NOT in the lead K. What is
     probably the purpose of this change?
   a. It eliminates the need for a NOT chip.
   b. It permits lead A to determine whether
      a shift or rotate will occur.
   c. It prevents a shift from occurring dur-
      ing an output.

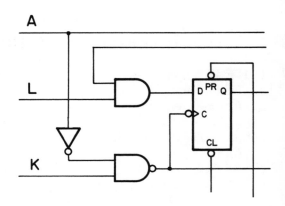

Name _____

Date _____ Score _____

Instructor _____

SHIFT REGISTER (7474, 2-7400 or 4013, 2-4011)

Shown is a universal shift register. Data can be input and output in either serial or parallel form. It differs from other circuits in that PR and CLR are not used to input parallel data.

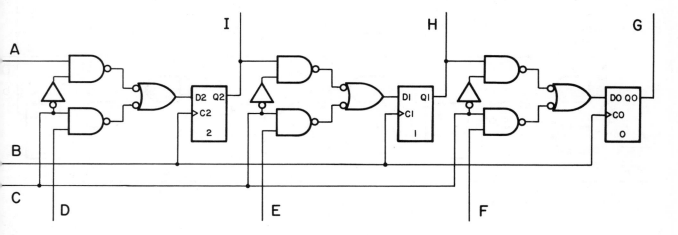

-1. Darken in (use a yellow marker or dark pencil) the path followed by a bit being shifted from Q2 to D1.

-2. To accomplish the shift described in Problem 1, what signal must be on control lead C?
   a. 1.
   b. 0.

-3. T   F   To accomplish the shift described in Problem 1, the clocks must be activated. Circle the correct answer.

-4. Darken in the path followed by a bit being input through parallel input F to DO.

-5. To accomplish the input described in Problem 4, what signal must be on control lead C?
   a. 1.
   b. 0.

-6. T   F   To accomplish the input described in Problem 4, the clocks must be activated. Circle the correct answer.

-7. Which notation would probably be used on control lead C?
   a. SHIFT/INPUT.
   b. INPUT/SHIFT.

-8. Laboratory Activity. If this is an experiment, build the first two bits of this circuit and test its action. Use NANDs for the NOTs.

Chapter 9
MASTER-SLAVE FLIP-FLOPS

Name _____

Date _____ Score _____

Instructor _____

SHIFT REGISTER APPLICATION (2-7474, 3-7400 or 2-4013, 3-4011)

The shift register shown can shift to either the right or the left.

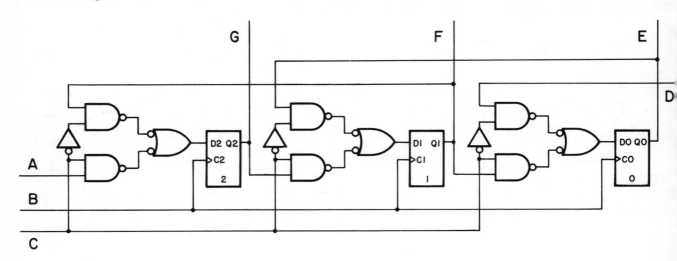

10-1. Darken in the path followed by a bit being shifted from Q2 (on flip-flop 2) to input D1 (o flip-flop 1).

10-2. To accomplish the shift to the right described in Problem 1, what signal must be on contr lead C?
   a. 1.
   b. 0.

10-3. Darken in the path followed by a bit being shifted from Q0 (on flip-flop 0) to D1 (on flip-flop 1

10-4. To accomplish the shift to the left described in Problem 3, what signal must be on control lead C
   a. 1.
   b. 0.

10-5. Which notation would be used on control lead C?
   a. RIGHT/LEFT.
   b. LEFT/RIGHT.

10-6. Match the following inputs and outputs with the letters on the leads of the above shift registe

   a. _____ Serial input when shifting right.

   b. _____ Serial input when shifting left.

   c. _____ Parallel outputs.

   d. _____ Serial output when shifting right.

   e. _____ Serial output when shifting left.

10-7. Laboratory Activity. If this is an experiment, build the circuit and test its action. Use NAND for NOTs.

Name _____

Date _____ Score _____

Instructor _____

FLIP-FLOPS

1-1. Match the following with the circuits at the right. When there are two spaces, supply two answers.

    a. _____, _____. Master-slave.

    b. _____, _____. Data latch.

    c. _____. Active-low, R-S inputs.

    d. _____. Edge-triggered clock.

    e. _____. Active-low level-triggered clock.

    f. _____. Active-high level-triggered clock.

    g. _____. Has preset and clear.

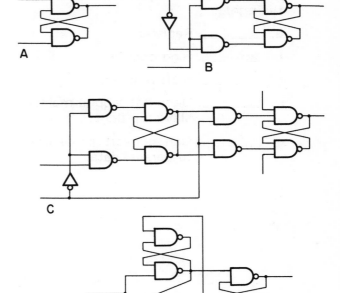

1-2. Match the following with the symbols at the right.

    a. _____. Has more than one output.

    b. _____. Active-high clock.

    c. _____. Edge-triggered clock.

    d. _____ and _____. Numbers can be entered without using clock.

1-3. Which graph best represents the output of this flip-flop?
    a. QA.
    b. QB.

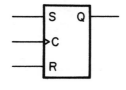

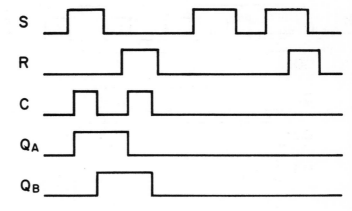

Chapter 9
MASTER-SLAVE FLIP-FLOPS

Name _____

Date _____ Score _____

Instructor _____

FLIP-FLOP APPLICATIONS

12-1. Match the following with the circuits at the right.

    a. _____. Register.

    b. _____. Ring counter.

    c. _____. Shift register.

    d. _____. Does not require master-slave flip-flops.

    e. _____. Can convert serial data to parallel data.

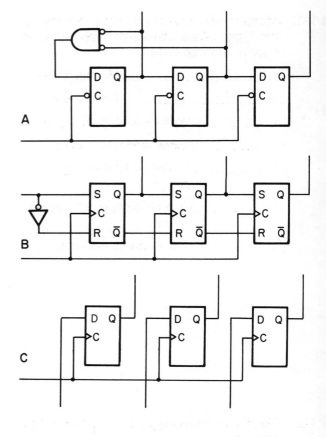

12-2. For the inputs shown, complete the timing diagram. Assume all outputs are 0 at t = 0.

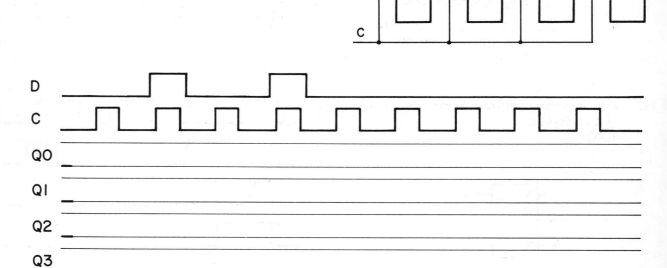

Chapter 10

TOGGLING FLIP-FLOPS

Name _____

Date _____ Score _____

Instructor _____

FLIP-FLOPS (2-7400, 7410 or 2-4011, 4023)

-1. Which truth table describes the action of a J-K flip-flop?

INPUTS A	B	OUTPUT Q_{t+1}
I	O	I
O	I	O
O	O	Q_t
I	I	$\overline{Q}_t$

A.

INPUTS A	B	OUTPUT Q_{t+1}
I	O	I
O	I	O
O	O	Q_t
I	I	*

B.

-2. Laboratory Activity. Shown is a circuit for a master-slave R-S flip-flop. Add the necessary leads to convert it to a J-K flip-flop. If this is an experiment, build this circuit and test its action against the truth table you selected in Problem 1.

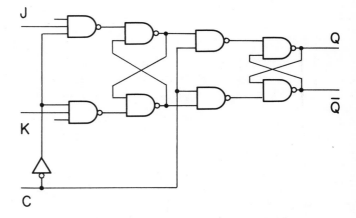

1-3. Match the following with the flip-flop symbols at the right. A given symbol may be used more than once.

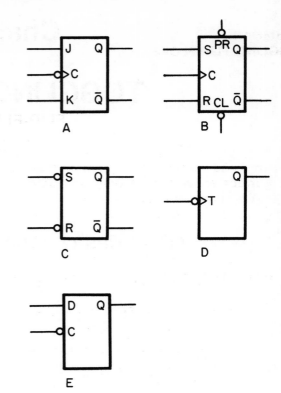

a. _____ and _____. These will toggle.

b. _____. Clock activated by positive-going edge.

c. _____. Level-triggered clock.

d. _____ and _____. Data can be input without clock signal.

e. _____. Has a set and a clear.

f. _____ and _____. These can be converted to a data latch by adding a NOT.

Chapter 10
TOGGLING FLIP-FLOPS

Name _____

Date _____ Score _____

Instructor _____

FLIP-FLOP CIRCUITS

2-1. Match the following with the circuit to the right and below. If there is more than one correct answer, give only one. A given circuit may be used more than once.

a. _____. Data latch.

b. _____. J-K flip-flop.

c. _____. Is not a flip-flop.

d. _____. A flip-flop without clock.

e. _____. Has preset and clear.

f. _____. Edge-triggered clock.

g. _____. Master-slave R-S flip-flop.

h. _____. Toggle flip-flop.

i. _____. Can toggle but is not called a toggle flip-flop.

j. _____. R-S flip-flop with active-low S and R inputs.

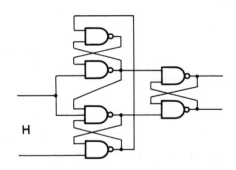

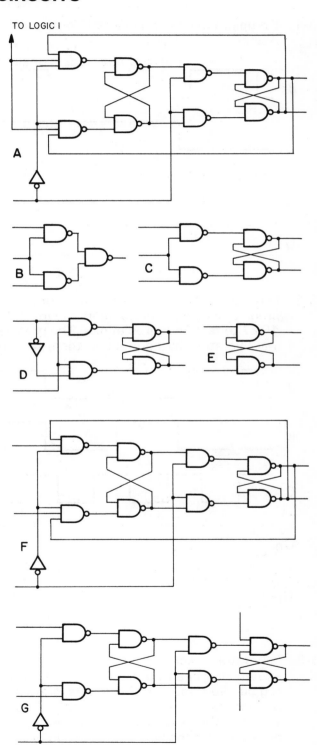

Chapter 10
TOGGLING FLIP-FLOPS

Name _____

Date _____ Score _____

Instructor _____

J-K FLIP-FLOP TIMING DIAGRAMS (7476 or 4027)

3-1. Complete this timing diagram for the J-K flip-flop shown. It is level-triggered and master-slave. Assume Q = 0 at t = 0.

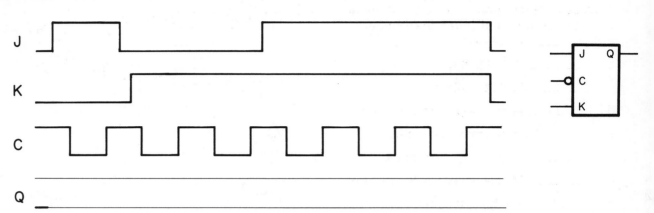

3-2. Laboratory Activity. Predict the output of this J-K flip-flop for the inputs shown. Record your predictions on the graph marked Qp. If this is an experiment, use one flip-flop from a 7476 or 4027 to test your predictions. The presets and clears on these flip-flops must be forced to the inactive state by connecting them to appropriate voltages.

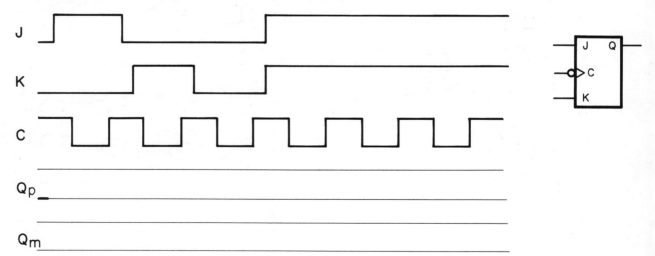

3-3. Laboratory Activity. The J-K flip-flop shown has been connected to act as a toggle flip-flop. For the inputs shown, predict its output. Assume Q = 0 at t = 0. If this is an experiment, test your predictions in the laboratory. Record your results on the graph marked Qm.

Chapter 10
TOGGLING FLIP-FLOPS

Name _____

Date _____ Score _____

Instructor _____

RIPPLE COUNTER (2-7476 or 2-4027)

4-1. Complete this diagram to show a 3-bit ripple counter. Include a lead that will permit clearing (setting all outputs to 0) the counter. Label this lead CLEAR. It is to be active-low.

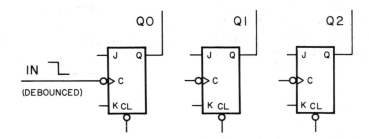

4-2. Using graphs Q0p, Q1p, and Q2p, predict the output of the above counter for the input shown. Assume all outputs are 0 at t = 0.

4-3. Laboratory Activity. If this is an experiment, build the above circuit. Experimentally test your predictions. Record your results in the unused graphs. Also test the action of CLEAR.

4-4. Laboratory Activity. Add a fourth bit (output Q3) to the above drawing. How many input pulses will be required for this 4-bit counter to go from 0000 to 0000? If this is an experiment, add the fourth bit to your circuit and determine its maximum count.

Predicted maximum count = _____ (base 10)   Measured maximum count = _____

Chapter 10
TOGGLING FLIP-FLOPS

Name _____

Date _____ Score _____

Instructor _____

DOWN COUNTER AND DEBOUNCING
(2-7476, 7400 or 2-4027, 4011, and 2-1K)

5-1. Laboratory Activity. Complete the timing
diagram for the circuit at the right. Complete the three graphs for the Q̄ outputs
first. Then construct the Q outputs by taking the complements of the Q̄s. Assume
all Q̄ outputs are 1 (all Q outputs are 0)
at t = 0.

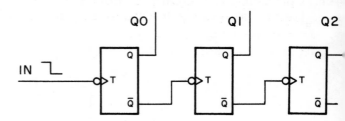

5-2. Laboratory Activity. Build the circuit and
test your predictions. Use a debounced
signal source for IN.

-3. Laboratory Activity. Replace the debounced signal source with a non-debounced source. Logic switches on most trainers are non-debounced. Observe the results of attempting to count signals from such a source.

-4. Laboratory Activity. Build the circuit shown and use its output to drive your down counter. Be sure the leads on the resistors are small enough in diameter (do not use resistors with power ratings of greater than 1/2 watt). If a SPDT switch is not available, use a grounded wire. Touch it to points 1 and 0 to input logic signals.

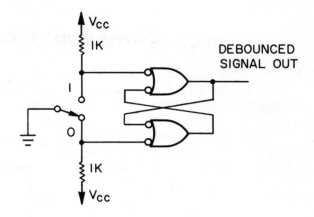

Name_____

Date _____ Score _____

Instructor _____

DECADE COUNTER (2-7476, 7400 or 2-4027, 4011)

6-1. Laboratory Activity. Complete the timing diagram for the following circuit. Include the spike
on Q1 and CLR that will be produced by the automatic reset process. Assume all outputs a
0 at t = 0.

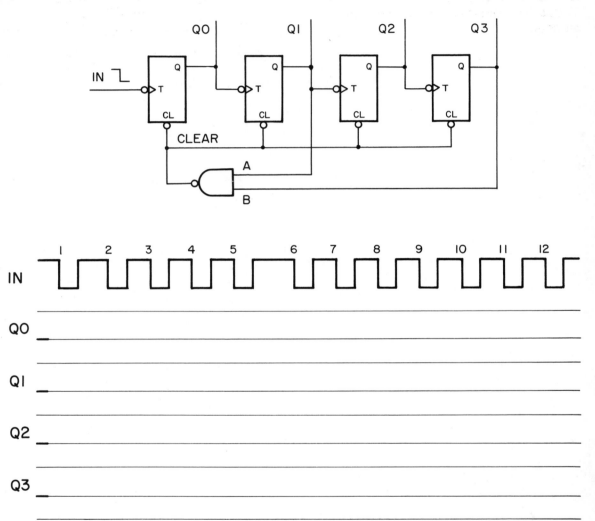

6-2. Laboratory Activity. Build the above circuit and test your predictions. If a logic probe capable of detecting pulses is available, it can be used to determine the presence of the spike at Q1. Count to 9 with the switch for input IN and place the probe on Q1. Pulse IN once more. You may be able to detect the spike at Q1 even though it is far too short to light the indicator lamp.

6-3. Laboratory Activity. Counters with various moduli can be obtained by moving leads A and B to different outputs. For each of the following connections, predict the maximum count. Then make the indicated connections and test your predictions.

a. A to Q0, B to Q3

Predicted count = _____. Measured count = _____.

b. A to Q2, B to Q3

Predicted count = _____. Measured count = _____.

c. A to Q1, B to Q2

Predicted count = _____. Measured count = _____.

Chapter 10

TOGGLING FLIP-FLOPS

Name _____

Date _____ Score _____

Instructor _____

ASYNCHRONOUS COUNTERS (2-7493)

7-1. Laboratory Activity. The following diagram shows the circuit within a 7493. Show how it cou
be connected to create a decade counter. Note that an external connection must be made betwe
the first and second flip-flop. Build the circuit and test its action.

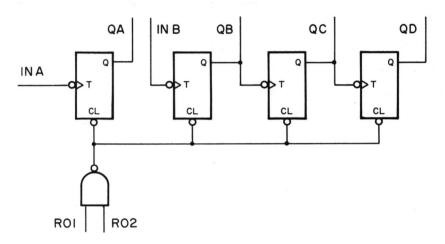

7-2. Laboratory Activity. Arrange with another laboratory group to use their decade counter a
tens counter. That is, the output of your counter is to drive the input of their counter. Conne
ONLY THE GROUND AND SIGNAL LEADS BETWEEN THE CIRCUITS. DO NOT CONNECT V
Count from 0 to at least 30 using the two counters.

7-3. Laboratory Activity. A scaller is a counter that outputs a pulse after a predetermined numk
of input pulses. Scallers are often used to reduce the amount of data sent on communicati
channels. Rather than send every pulse, only every fifth, tenth, or hundredth pulse is sen

Show how a 7493 could be connected to output a negative-going edge for every five inp
pulses. Then connect the scaller to your decade counter and demonstrate its action.

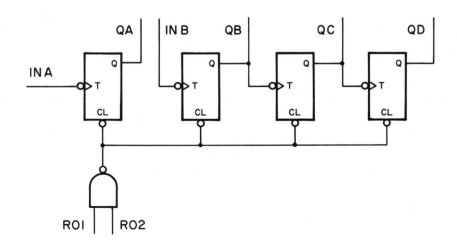

Chapter 10
TOGGLING FLIP-FLOPS

Name _____

Date _____ Score _____

Instructor _____

PRESET COUNTER (2-7476 or 2-4027)

10-1. Complete the timing diagram for the circuit at the right. The number $Q2Q1Q0 = 101$ has been preset.

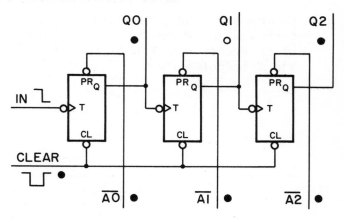

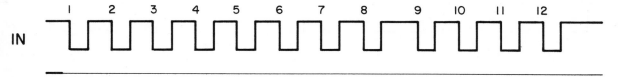

10-2. After the above counter has reached 000, what number will it count to if a new preset is not entered? N = _____ (base 10).

10-3. Indicate the preset that would be entered to obtain each of the following maximum counts.
   a. For 5 counts to 000:   Q2 = _____.   Q1 = _____.   Q0 = _____.
   b. For 7 counts to 000:   Q2 = _____.   Q1 = _____.   Q0 = _____.

10-4. Laboratory Activity. If this is an experiment, build the above circuit and test its action. Use the logic switches on your trainer for CLEAR, A0, A1, and A2. These switches do not have to be debounced. The switch at IN, however, must be debounced. If 7476s are used, PR and CL are active-low, so these inputs must be high for the circuit to count. SET and RESET on the 4027 are active-high. These inputs must be low for the circuit to count.

   a. Predicted count for $Q2Q1Q0 = 101$ is _____.   Measured count is _____.

   b. Predicted count for $Q2Q1Q0 = 011$ is _____.   Measured count is _____.

   c. Predicted count for $Q2Q1Q0 = 001$ is _____.   Measured count is _____.

Chapter 10

TOGGLING FLIP-FLOPS

Name _____

Date _____ Score _____

Instructor _____

FREQUENCY DIVIDER (2-7476 or 2-4027)

9-1. Complete the timing diagram for the frequency divider at the right. At t = 0, the circuit reached its maximum count of 16, and all outputs went to 0.

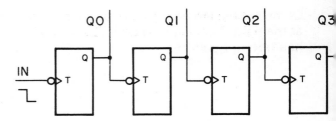

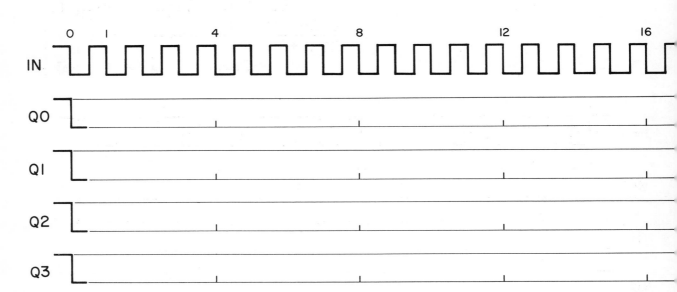

9-2. Based on the above timing diagram, determine the number of cycles at IN required to produce one cycle at each of the following outputs. Also determine the ratio of the frequency at each output to the frequency at IN.

    a. 1 cycle at Q0 requires _____ cycles at IN. $f(Q0)/f(IN) = 1/$_____.

    b. 1 cycle at Q1 requires _____ cycles at IN. $f(Q1)/f(IN) = 1/$_____.

    c. 1 cycle at Q2 requires _____ cycles at IN. $f(Q2)/f(IN) = 1/$_____.

    d. 1 cycle at Q3 requires _____ cycles at IN. $f(Q3)/f(IN) = 1/$_____.

9-3. Laboratory Activity. If this is an experiment, build the above circuit and test your predictions. If available, a frequency counter can be used. If not, set the clock source on your trainer to a low frequency (you may have to add an external capacitor). Attempt to count the cycles at the various outputs. Record your results below.

    a. Measured $f(Q0)/f(IN) = 1/$_____.

    b. Measures $f(Q1)/f(IN) = 1/$_____.

    c. Measured $f(Q2)/f(IN) = 1/$_____.

    d. Measured $f(Q3)/f(IN) = 1/$_____.

Name _____

Date _____ Score _____

Instructor _____

FLIP-FLOP APPLICATIONS

10-1. Match the following with the circuits shown. When there is more than one correct answer, give only one. A given letter may be used more than once.

a. _____.Register.

b. _____.Shift register.

c. _____.Ring counter.

d. _____.Converts serial data to parallel data.

e. _____.Does not require master-slave.

f. _____.Ripple counter.

g. _____.Down counter.

h. _____.Counter with debouncer.

i. _____.Counts to 8.

j. _____.Synchronous counter.

k. _____.Divides frequency by 8.

l. _____.Decade counter.

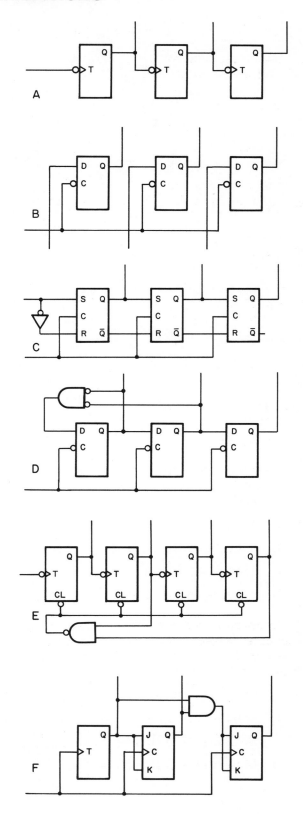

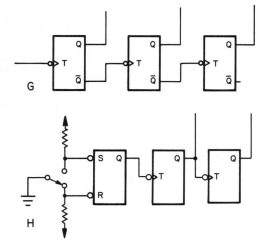

Chapter 11

BINARY NUMBER SHORTCUTS AND ARITHMETIC PROCESSING METHODS

Name _____

Date _____ Score _____

Instructor _____

BINARY-TO-DECIMAL AND DECIMAL-TO-BINARY CONVERSION

1-1. Complete the table shown by counting to 23 in binary.

1-2. Based on the binary number

   11,001,100

   indicate the values (1s and 0s) of the following bits.

   a. D1 = _____.

   b. D2 = _____.

   c. D3 = _____.

1-3. Indicate the place values of the following bits of a binary number.

   D7 D6 D5 D4 D3 D2 D1 D0

   —   —   —   —   —   —   —   —

1-4. Convert the following binary numbers to decimal.

	Base 2	Base 10
a.	1,100	_____
b.	111,111	_____
c.	1,001,010	_____
d.	10,000,000,000	_____

1-5. Convert the following decimal numbers to binary. Write the binary bits in groups of three.

	Base 10	Base 2
a.	24	_____
b.	82	_____
c.	286	_____

DECIMAL	BINARY
0	
1	
2	
3	
4	
5	
6	
7	
8	
9	
10	
11	
12	
13	
14	
15	
16	
17	
18	
19	
20	
21	
22	
23	

145

Chapter 11

**BINARY NUMBER SHORTCUTS AND
ARITHMETIC PROCESSING METHODS**

Name _____

Date _____ Score _____

Instructor _____

UNSIGNED BINARY ADDITION AND OVERFLOW

2-1. Do the following additions. Do not be concerned about overflow.

   a.   110
    + 11

   b.   11,100
    +   111

   c.   101,110,110
    +   111,100

2-2. Show how the following decimal numbers would be represented as unsigned numbers in an 8-bit computer. If a number is too large for an 8-bit word, mark it TOO LARGE.

   a. 300 (base 10) = ____ ____,____ ____ ____,____ ____ ____ (base 2)

   b. 102 (base 10) = ____ ____,____ ____ ____,____ ____ ____ (base 2)

   c.   44 (base 10) = ____ ____,____ ____ ____,____ ____ ____ (base 2)

2-3. Add the following unsigned numbers. In addition to the 8-bit sum, indicate whether or not the carry flag would be set (1) or cleared (0). If an overflow results, circle the letter (a, b, c, d, e, f) in front of the numbers.

   a.   00,010,000
    + 01,110,011

    c__ _____

   b.   11,001,100
    + 00,101,000

    c__ _____

   c.   01,001,010
    + 10,111,000

    c__ _____

   d.   10,001,000
    + 10,001,000

    c__ _____

   e.   00,010,001
    + 10,110,000

    c__ _____

   f.   01,111,111
    + 01,111,111

    c__ _____

2-4. The following are double-precision numbers. When added, the carry from the lower-order byte is automatically added to the LSB of the higher-order byte. Repeat Problem 3 for these numbers.

             c__

   a.   01,010,111   01,100,000
    + 01,011,000   11,001,000

    c__ _____   _____

             c__

   b.   10,000,100   01,111,111
    + 01,111,100   11,000,000

    c__ _____   _____

Chapter 11
BINARY NUMBER SHORTCUTS AND
ARITHMETIC PROCESSING METHODS

Name _____

Date _____ Score _____

Instructor _____

TWO'S COMPLEMENT AND SIGNED NUMBERS

11-1. Write the two's complement of these numbers. The number at d is double-precision.
   a. 00,010,111
   b. 11,010,111
   c. 00,100,000
   d. 00,010,011   11,010,110

11-2. Use two's complement to accomplish the following subtractions. These are unsigned numbers. Draw a box around your results. Ignore overflow. Show all of your work in the spaces provided.

   a. 01,011,011 − 00,011,101

   b. 11,011,000 − 01,100,100

11-3. Write P (positive) or N (negative) in front of each of the following signed numbers. The number at d is from a 16-bit machine.

   a. _____ 00,001,100

   b. _____ 01,111,110

   c. _____ 10,001,100

   d. _____ 1100,0110,1000,0011

11-4. Indicate the largest number that can be represented by each of the following registers when unsigned and signed numbers are stored.

	Unsigned	Signed
8-bit register	_____ (base 10)	_____ (base 10)
16-bit register	_____ (base 10)	_____ (base 10)

Chapter 11
BINARY NUMBER SHORTCUTS AND
ARITHMETIC PROCESSING METHODS

Name _____

Date _____ Score _____

Instructor _____

SIGNED ADDITION

4-1. Determine the decimal values of these signed binary numbers. Be sure to include the signs

    a. 00,011,101 = _____ (base 10)

    b. 11,001,001 = _____ (base 10)

    c. 01,110,000 = _____ (base 10)

    d. 11,111,000 = _____ (base 10)

4-2. Do the following additions. Indicate overflow by circling the letter (a, b, c, d, e, f) in front of the numbers that result in overflow. Carry does not need to be considered, since these are signed numbers.

    a.   00,100,101
        + 10,100,100

    b.   10,011,100
        + 01,110,000

    c.   01,001,111
        + 00,011,000

    d.   00,100,010
        + 01,011,110

    e.   11,001,000
        + 11,011,000

    f.   10,110,000
        + 11,100,000

4-3. Repeat Problem 2 for this double-precision number. Take the carry into account.

                     c__
    00,100,100   10,110,000
  + 00,110,000   11,001,000

Chapter 11
BINARY NUMBER SHORTCUTS AND
ARITHMETIC PROCESSING METHODS

Name _____

Date _____ Score _____

Instructor _____

OVERFLOW

The rules of overflow for signed binary numbers are:

    a. Overflow will not occur if the sign bits of the two numbers to be added differ (one is positive, one is negative).

    b. Overflow has occurred when the sum of two numbers with the same sign (both positive or both negative) has the opposite sign.

5-1. Using the above rules, complete the truth table shown. OV stands for OVERFLOW. OV = 1 implies an overflow.

    augend   S1X,XXX,XXX

    addend   S2X,XXX,XXX

    Sum   S3X,XXX,XXX

5-2. Using the sum-of-products method, write the Boolean expression represented by this truth table. Do not simplify.

    _____ = OV

SIGN BIT			
AUGEND	ADDEND	SUM	OVERFLOW
S1	S2	S3	OV
0	0	0	
0	0	1	
0	1	0	
0	1	1	
1	0	0	
1	0	1	
1	1	0	
1	1	1	

5-3. Some computers have overflow flags. When an arithmetic operation produces an overflow, a flip-flop is set (Q = 1). Such flags can be tested to see if the results of a computation are valid.

To show a simple version of such a circuit, implement the above expression. Because the three sign bits do not exist at the same time, flip-flops have been provided to store these bits. Note that Q and $\bar{Q}$ are available for S1, S2, and S3.

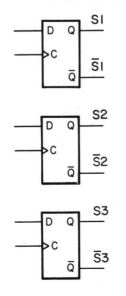

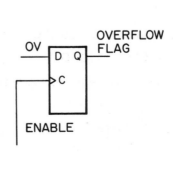

Chapter 11
BINARY NUMBER SHORTCUTS AND
ARITHMETIC PROCESSING METHODS

Name _____

Date _____ Score _____

Instructor _____

REVIEW

6-1. Convert these binary numbers to base 10.

a. 00,011,011 = _____ (base 10)
b. 10,001,000 = _____ (base 10)

6-2. Complete this equation for expressing binary numbers in terms of base 10 numbers. Exten it to bits D2, D3, and D4.

_____ + _____ + _____ + D1(2) + D0(1) = N (base 10)

6-3. Using the MSB method, convert this base 10 number to binary.

20 (base 10) = _____ (base 2)

6-4. Using the LSB method, convert this base 10 number to binary.

13 (base 10) = _____ (base 2)

6-5. Add these unsigned, 8-bit binary numbers. If an overflow results, circle your answer.

a.   10,100,001
   + 01,100,010

b.   11,100,001
   + 00,011,100

6-6. Add these signed, 8-bit binary numbers. If an overflow results, circle your answer.

a.   00,010,100
   + 01,110,000

b.   11,011,000
   + 01,001,000

6-7. Convert these signed, binary number to base 10. Include their signs.

a. 01,000,010 = _____ (base 10)
b. 10,111,010 = _____ (base 10)

6-8. Use two's-complement to do the following subtraction. These are unsigned numbers.

01,010,100 − 00,111,000 = _____ (base 2)

Check your answer in base 10 with the following table:

    (   ) (base 10)
− (   ) (base 10)
    (   ) (base 10)

Chapter 12

NUMBERING SYSTEMS AND CODES OTHER THAN BINARY SYSTEM

Name ____________________

Date ____________ Score ____________

Instructor ____________________

COUNTING USING VARIOUS BASES

1-1. Complete the table shown. Each column represents a different numbering system.

DECIMAL BASE 10	OCTAL BASE 8	HEXADECIMAL BASE 16	DUODECIMAL BASE 12	BINARY BASE 2
0	0	0	0	0
1	1	1	1	1
2	2	2	2	10
3	3	3	3	
4	4	4	4	
5	5	5	5	
6		6	6	
7		7	7	
8		8	8	
9		9	9	
10		A	A	
11	13	B	B	
12				
13				
14	16	E		
15	17			
16				
17		11		
18			16	
19				
20				
21				
22				
23				
24		18		
25				
26	32		22	11010

Chapter 12
NUMBERING SYSTEMS AND CODES
OTHER THAN BINARY SYSTEM

Name _____

Date _____ Score _____

Instructor _____

OCTAL AND OTHER NUMBERING SYSTEMS

2-1. Indicate the bases of these numbering systems. Express your answers in decimal.

0	0	0	0	0
1	1	1	1	1
2	2	2	10	2
3	3	3	11	3
4	4	4	100	4
10	5	5	101	5
11	6	6	110	6
12	7	7	111	7
13	10	8	1000	8
14	11	9	1001	9
20	12	10	1010	A
21	13	11	1011	B
22	14	12	1100	C
23	15	13	1101	D
24	16	14	1110	10
30	17	15	1111	11
31	20	16	10000	12

a. _____   b. _____   c. _____   d. _____   e. _____

2-2. Which of the following numbers could not be octal?
a. 101.
b. 572.
c. 89.

2-3. Convert these binary numbers into octal.

a.   100,110 = _____ (base 8).

b.   001,101 = _____ (base 8).

c. 100,011,111 = _____ (base 8).

d.   11,000 = _____ (base 8).

2-4. Convert these octal numbers to binary. Omit leading zeros.

a.   47 = _____ (base 2).

b.   77 = _____ (base 2).

c. 701 = _____ (base 2).

d. 352 = _____ (base 2).

2-5. Indicate the numbers represented by these monitor lamps. A solid dot represents a 1; an ope
dot a 0. Give your answer in octal.

a. _____ (base 8).

○ ○ ●   ○ ● ●   ● ○ ○

b. _____ (base 8).

● ○ ○ ● ○ ○ ●

**Chapter 12
NUMBERING SYSTEMS AND CODES
OTHER THAN BINARY SYSTEM**

Name _____

Date _____ Score _____

Instructor _____

OCTAL NUMBERING SYSTEM

12-1. Convert these octal numbers to decimal. Show your work.

    a. 63 (base 8) = _____(base 10).   b. 372 (base 8) = _____(base 10).

12-2. Convert these decimal numbers to octal. Use the MSD method. Show your work.

    a. 424 (base 10) = _____(base 8).   b. 250 (base 10) = _____(base 8).

12-3. Convert these decimal numbers to octal. Use the LSD method. Show your work.

    a. 703 (base 10) = _____(base 8).   b. 492 (base 10) = _____(base 8).

12-4. For each pair of numbers, circle the one representing the larger count.

    a. 132 (base 8) or 130 (base 10).

    b. 111 (base 8) or 111 (base 10).

Chapter 12
NUMBERING SYSTEMS AND CODES
OTHER THAN BINARY SYSTEM

Name _____

Date _____ Score _____

Instructor _____

HEXADECIMAL NUMBERING SYSTEM

4-1. Convert these binary numbers to hexadecimal.

    a.   1000,1001 = _____ (base 16).

    b.   0111,1011 = _____ (base 16).

    c.   1100,1010 = _____ (base 16).

    d. 0110,1101,1111 = _____ (base 16).

4-2. Indicate the numbers represented by these monitor lamps. Give your answers in both hexadecim
and octal.

    a. _____ (base 16), _____ (base 8).

    b. _____ (base 16), _____ (base 8).

    c. _____ (base 16), _____ (base 8).

    ● ○ ● ○   ● ● ○ ●

    ○ ●   ● ○ ○   ● ● ●

    ○ ● ● ○ ● ○ ○ ○ ● ● ○ ○

4-3. Convert these hexadecimal numbers to binary. Do not show leading zeros.

    a. 92 = _____ (base 2).

    b. A37 = _____ (base 2).

    c. ADF = _____ (base 2).

    d. DOB = _____ (base 2).

4-4. Make the following conversions by changing the original number to binary and then converti
it to the indicated base.

    a. 452 (base 8) = _____ = _____ (base 16).

    b. A4 (base 16) = _____ = _____ (base 8).

4-5. For each pair of numbers, circle the one representing the larger count. If they are equal,
circle both.

    a.   D (base 16) or 10 (base 10).

    b. 111 (base 2) or 111 (base 16).

    c.   77 (base 8) or 77 (base 16).

    d. 100 (base 2) or 10 (base 8).

    e.   11 (base 10) or 111 (base 2).

4-6. This double-precision number is stored in two memory locations in an 8-bit computer. Expre
it in both octal and hexadecimal.

    a. N = _____ (base 16).   High-order byte   Low-order byte

    b. N = _____ (base 8).   00110100   10100010

Chapter 12
NUMBERING SYSTEMS AND CODES
OTHER THAN BINARY SYSTEM

Name _____

Date _____ Score _____

Instructor _____

HEXADECIMAL NUMBERING SYSTEM

1. Indicate the place values of the following digits of hexadecimal numbers.

        D 5   D 4   D 3   D 2   D 1   D 0

        _____   _____   _____   _____   _____   _____

2. Convert these hexadecimal numbers to decimal. Show your work.

   a. A5 = _____ (base 10).   b. FAC = _____ (base 10).

3. Convert these decimal numbers to hexadecimal. Use the MSD method. Show your work.

   a. 300 (base 10) = _____ (base 16).   b. 687 (base 10) = _____ (base 16).

4. Convert these decimal numbers to hexadecimal using the LSD method. Show your work.

   a. 1022 (base 10) = _____ (base 16).   b. 2574 (base 10) = _____ (base 16).

Chapter 12
NUMBERING SYSTEMS AND CODES
OTHER THAN BINARY SYSTEM

Name _____

Date _____ Score _____

Instructor _____

BINARY-CODED-DECIMAL NUMBERING SYSTEM

	A	B
	0	0
	1	1
	10	10
	11	11
	100	100
	101	101
	110	110
	111	111
	1000	1000
	1001	1001
	10000	1010
	10001	1011
	10010	1100
	10011	1101
	10100	1110
	10101	1111
	10110	10000

6-1. Which column at the right represents BCD?
   a. Column A.
   b. Column B.
   c. Both are BCD.

6-2. Convert these BCD numbers to decimal.
   a. 0110,1000 = _____ (base 10).
   b. 0011,1001 = _____ (base 10).
   c. 1000,0100 = _____ (base 10).

6-3. Convert these decimal numbers to BCD.
   a. 52 = _____(BCD).
   b. 293 = _____(BCD).
   c. 28 = _____(BCD).

6-4. Which number could not be BCD?
   a. 0001,1001.
   b. 0111,0111.
   c. 0100,1100.

6-5. Which number represents the higher count?
   a. 1000,0111 (BCD).
   b. 1000,0111 (base 2).

6-6. Show how two 7493 counter chips could be connected to form a 2-decade counter with BCD output. Do not be concerned that the outputs are in the reverse order. This would be correct when indicator lamps are added. Note the location of IN.

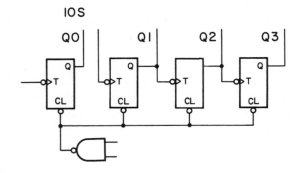

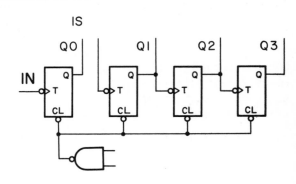

Chapter 12
NUMBERING SYSTEMS AND CODES
OTHER THAN BINARY SYSTEM

Name _____

Date _____ Score _____

Instructor _____

NUMBERING SYSTEMS

-1. At the right, a solid dot represents an activated lamp; an open dot represents an inactive lamp. Record the indicated number in each of the following forms.

a. _____ (binary).

b. _____ (octal).

c. _____ (hexadecimal).

d. _____ (decimal from BCD).

● ○ ● ○ ○ ● ● ○

-2. Which set of lamps displays the hexadecimal number BE?

a. ○ ● ● ●   ● ○ ○ ●

b. ● ○ ● ●   ● ● ● ○

c. ○ ● ○ ●   ○ ● ○ ○

-3. By drawing in, show how these lamps would indicate the hexadecimal number 4D. An open dot represents a lamp that is out (logic 0). A solid dot represents a lamp that is lit (logic 1).

○ ○ ○ ○   ○ ○ ○ ○

-4. This two-digit, 7-segment display indicates a hexadecimal number. Convert it to decimal.

N = _____ (base 10).

-5. This card is used to input programs and data to a small computer. Numbers must be in binary. The rectangles are indented so they may be hand punched. A hole (shown as a dark rectangle) represents a 1.

To increase readability, numbers are arranged in groups of three. This is called BCO (binary-coded-octal). Indicate the BCO numbers represented by rows a and b.

a. Na = _____ (base 8).

b. Nb = _____ (base 8).

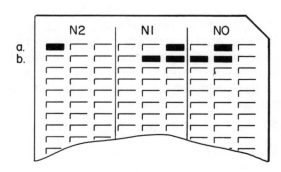

Chapter 12

NUMBERING SYSTEMS AND CODES
OTHER THAN BINARY SYSTEM

Name _____

Date _____ Score _____

Instructor _____

CODES AND NUMBERS

8-1. Which column represents the gray code?
 a. Column A.
 b. Column B.
 c. Both are gray code.

A	B
0	0
1	1
11	10
10	11
110	100
111	101
101	110
100	111
1100	1000

8-2. Approximately how many words in memory are needed to store the following in ASCII in a 8-bit machine? WHAT IS YOUR NAME?
 a. 1 word.
 b. 4 words.
 c. 18 words.

8-3. Match the following terms with the parts of the number at the right.

 a. _____ Byte.

 b. _____ Bit.

 c. _____ Word.

A

10011001 01001101

B

C

8-4. The following represents BCD numbers stored in a computer. Which obviously represents a packed word?
 a. 0011,1001.
 b. 0000,1001.

8-5. This integrated circuit is a BCD-to-decimal decoder. For the input shown, circle the output that will be active (low).

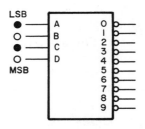

8-6. The starred elements are decoder-drivers. They accept binary numbers at their inputs and drive the segments of the 7-segment displays. For the hexadecimal numbers being output, what binary number is at the inputs of the decoders?

A7 A6 A5 A4 A3 A2 A1 A0

— — — — — — — —

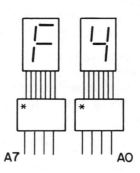

Chapter 13

ARITHMETIC/LOGIC CIRCUITS

Name _____

Date _____ Score _____

Instructor _____

MULTI-BIT ADDER (7400, 7408, 7486 or 4011, 4070, 4081)

-1. Complete this truth table to show the action of a full adder.

-2. Using sum-of-products and the truth table from Problem 1, write the expression for S and Co of a half adder.

_____ = S

_____ = Co

INPUTS			PREDICTED OUTPUTS		MEAS. OUTPUTS	
A	B	C	S	Co	S	Co
0	0	0				
0	0	1				
0	1	0				
0	1	1				
1	0	0				
1	0	1				
1	1	0				
1	1	1				

-3. Laboratory Activity. The circuit at the right is a full adder. For the input set shown, predict the signals at the indicated points. Circle this input set in the above truth table to indicate that your results match the predicted results. If this is an experiment, build the full adder and verify the column of the truth table from Problem 1 marked PREDICTED. Write your results in the column marked MEAS.

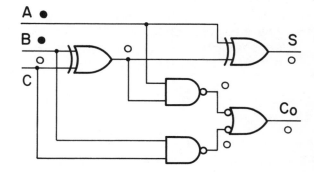

1-4. Laboratory Activity. For each of the following input sets, predict the output of the 2-bit add shown. Record your predictions in the rows marked Predic. If this is an experiment, build t half adder shown. Use it and the circuit from Problem 3 to test your predictions. The half add circuit takes care of the A0 and B0 inputs. The 1-bit full adder takes care of A1, B1, and C

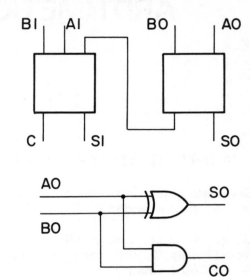

a. A1A0   =   01
    B1B0   =   + 10

    Predic. sum = c__ _____

    Meas. sum   = c__ _____

b. A1A0   =   11
    B1B0   =   + 11

    Predic. sum = c__ _____

    Meas. sum   = c__ _____

c. A1A0   =   10
    B1B0   =   + 10

    Predic. sum = c__ _____

    Meas. sum   = c__ _____

Chapter 13
ARITHMETIC/LOGIC CIRCUITS

Name _____

Date _____ Score _____

Instructor _____

MULTI-BIT ADDER CIRCUIT

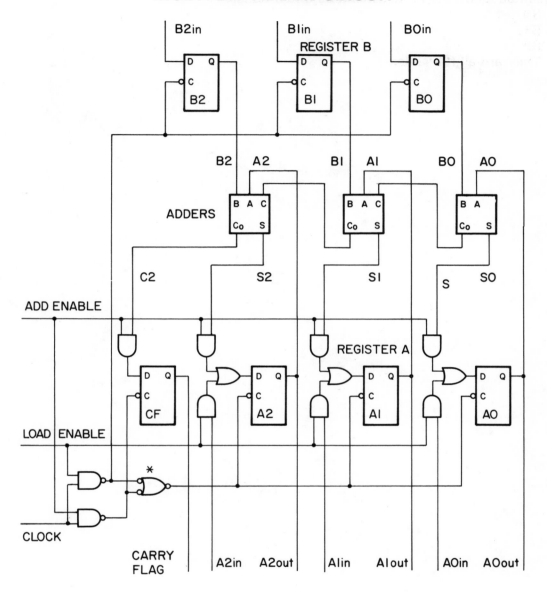

A 3-bit adder is shown. Numbers to be added are placed in Registers A and B. Their sum is returned to Register A. Questions on this and the next sheet refer to this circuit.

2-1. Which best describes the clock inputs on the flip-flops?
    a. Negative-going, edge-triggered.
    b. Positive-going, edge-triggered.
    c. Active-low, level-triggered.
    d. Active-high, level-triggered.

2-2. What is the starred element at the lower left?
   a. AND.
   b. NAND.
   c. OR.
   d. NOR.

2-3. Which adder would be described as a half adder?
   a. S0.
   b. S1.
   c. S2.
   d. They are all full adders.

Chapter 13
ARITHMETIC/LOGIC CIRCUITS

Name _____

Date _____ Score _____

Instructor _____

MULTI-BIT ADDER CIRCUIT

Problems on this sheet refer to the circuit shown for Problem 2-1 through Problem 2-3 (called 13-2-1 through 13-2-3).

3-1. Darken in the path of a bit from input A2in, through the flip-flop, to input A2 of the adder.

3-2. To permit a number from input A2in to be loaded into flip-flop A2, what number must be placed on control lead LOAD ENABLE?
   a. 1.
   b. 0.

For True or False questions, circle the correct answer.

3-3.   T   F   To load a number from the A inputs into Register A, the clocks on the flip-flops must be activated.

3-4. Darken in the path of a bit from S1 (the output of the second adder), through the associated flip-flop, to output A1out. Also, heavy in the path of this bit back to input A1 at the adder.

3-5. To permit an addition, what number must be placed on control lead ADD ENABLE?
   a. 1.
   b. 0.

3-6.   T   F   To complete the storage of the sum in Register A, the clocks on the flip-flops of Register A must be activated.

3-7.   T   F   Because the outputs of the flip-flops on Register A are returned to the adders, these flip-flops must be master-slave.

3-8.   T   F   The flip-flops in Register B must also be master-slave.

3-9.   T   F   As soon as numbers are stored in Registers A and B, their sum appears at the outputs of the three adders.

3-10. The circuit driving the clock of flip-flop CF differs from that of the clocks in Register A. Which of the following best describes the storing of a new number in flip-flop CF?
   a. A new number enters CF only during the storing of a new sum.
   b. A new number enters CF during both the storing of a new sum and when a new number enters through input A.

3-11. Assume Register A = 100 and Register B = 001. What number will appear in Register A and CF after ADD ENABLE and CLOCK have been activated?

   CF = _____.   Register A = _____.

3-12. Refer to Problem 11 (called Problem 3-11 or 13-3-11). If ADD ENABLE and CLOCK were activated a second time (without placing a new number in Register A), what numbers would appear in CF and Register A?

   CF = _____.   Register A = _____.

Chapter 13
ARITHMETIC/LOGIC CIRCUITS

Name _____

Date _____ Score _____

Instructor _____

ADDER/SUBTRACTOR (7486, 7483 or 4070, 4008)

4-1. Laboratory Activity. Do the following 4-bit additions. Record your answers in the row marked Predicted sums. These are unsigned numbers.

Number A	a.	1100	b.	1100	c.	1100
Number B		+ 0010		+ 0101		+ 0011

Predicted sums c__ _____ c__ _____ c__ _____

Measured sums c__ _____ c__ _____ c__ _____

4-2. Laboratory Activity. Using two's complement, do the following subtractions. Record your answers in the row marked Predicted diff.

Number A	a.	1100	b.	1100	c.	1100
Number B		− 0010		− 0101		− 0011

$\overline{2}$s of B
Number A   + 1100   + 1100   + 1100

Predicted dif. c__ _____ c__ _____ c__ _____

Measured dif. c__ _____ c__ _____ c__ _____

4-3. Laboratory Activity. Build the circuit shown and experimentally test your predictions. Record your results in the rows marked Measured sum and Measured dif. If your trainer has only four switches, number A can be obtained by connecting A3 and A2 to Vcc and A1 and A0 to ground. Also, if your trainer has only four lamps, connect a long lead to lamp S3 and move it between S3 and carry out.

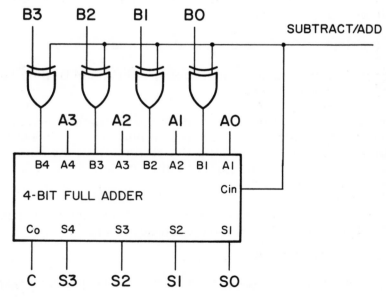

Chapter 13
ARITHMETIC/LOGIC CIRCUITS

Name _____

Date _____ Score _____

Instructor _____

COMPARATOR (7402, 7486, 7400, 7410 or 4001, 4070, 4011, 4023)

5-1. Laboratory Activity. The output of the comparator shown should be 1 when A1A0 equals B1B0. For all other input sets, its output should be 0. For the inputs shown, use truth table circuit analysis at the indicated points. At the output, use only the upper dot. If this is an experiment, build the circuit and test your predictions. Use the lower dot at the circuit's output to record your results.

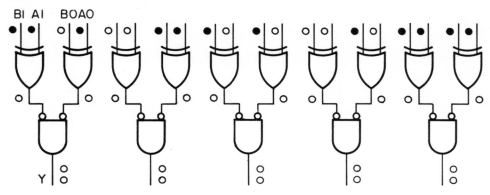

5-2. The circuits shown output 1s when A1A0 is larger than B1B0. For the inputs shown, use truth table circuit analysis to determine the signals at the indicated points.

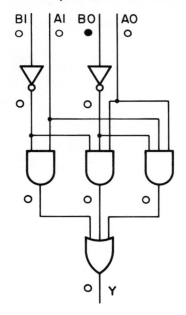 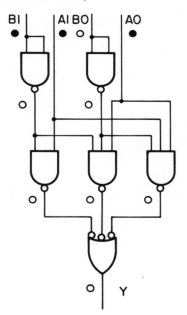

5-3. Laboratory Activity. Based on the action of the above circuits, complete the PREDICTED OUTPUT column of this truth table. If this is an experiment, build the circuit at the right above and test your predictions. Write your results in the column marked MEAS. OUTPUT. Remember, the circuit's output equals 1 only when A1A0 is larger than B1B0.

INPUTS				PREDICTED OUTPUT A > B	MEAS. OUTPUT A > B
A_1	A_0	B_1	B_0		
0	0	0	0		
1	0	0	1		
1	1	0	1		
0	1	1	0		
1	0	1	1		

**Chapter 13
ARITHMETIC/LOGIC CIRCUITS**

Name _____

Date _____ Score _____

Instructor _____

SHIFT/ROTATE REGISTER

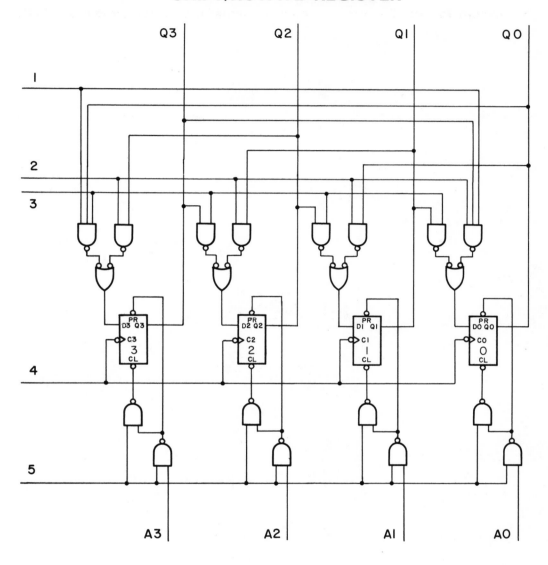

The register shown can shift or rotate to either the right or left. By answering a series of questions (on this sheet and the next), you will analyze its action.

6-1. Parallel numbers enter through A3A2A1A0. Which control lead would be labeled LOAD ENABLE?
   a. 1.
   b. 2.
   c. 3.
   d. 4.
   e. 5.

For True or False questions, circle the correct answer.

6-2. T   F   LOAD ENABLE is active-high.

6-3. T   F   To input a parallel number through A3A2A1A0, the clocks on the flip-flops must b
            activated.

6-4. T   F   Except for the flip-flops, all logic elements used in the circuit are NANDs.

Chapter 13
ARITHMETIC/LOGIC CIRCUITS

Name _____

Date _____ Score _____

Instructor _____

SHIFT/ROTATE REGISTER

Problems on this sheet refer to the circuit shown for Problem 6-1 through Problem 6-4 (called 13-6-1 through 13-6-4).

-1. Start at output Q3 of data latch 3 and darken in the signal path to input D2 of data latch 2. This is the path followed by data being shifted to the right.

-2. Based on the path indicated in Problem 1, which control lead would be labeled RIGHT ENABLE?
 a. 1.
 b. 2.
 c. 3.
 d. 4.
 e. 5.

Circle the correct answer.

-3. T   F   Control lead RIGHT ENABLE is active-high.

-4. T   F   To shift or rotate to the right, the clocks of the flip-flops must be activated.

-5. Start at output Q1 on data latch 1 and darken in the signal path to input D2 on data latch 2. This is the path followed by data being shifted to the left.

-6. Based on the path indicated in Problem 5, which control lead would be labeled LEFT ENABLE?
 a. 1.
 b. 2.
 c. 3.
 d. 4.
 e. 5.

-7. T   F   Control lead LEFT ENABLE is active-high.

-8. Use a series of arrows to show the signal path between output Q0 of data latch 0 to input D3 of data latch 3. This is the return path used during a rotate right.

-9. Control lead 1 determines whether a shift or rotate will be performed. Note that it is connected only to the two 3-input NANDs. Which notation would be used on this lead?
 a. SHIFT/ROTATE.
 b. ROTATE/SHIFT.

-10. Label the control leads on this block diagram of the circuit on the sheet with 13-6-1 through 13-6-4. Lead 4, the CLOCK, has already been labeled.

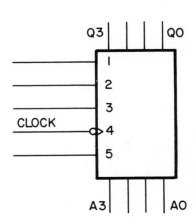

7-11. Indicate the control signals (1s and 0s) needed to do a shift to the right. CLOCK would be a negative-going edge.

    a. Lead 1 = _____.

    b. Lead 2 = _____.

    c. Lead 3 = _____.

    d. Lead 5 = _____.

Chapter 13
ARITHMETIC/LOGIC CIRCUITS

Name _____

Date _____ Score _____

Instructor _____

ANDing CIRCUIT

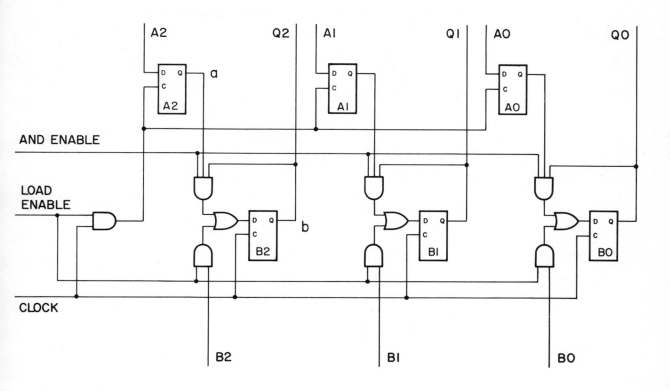

Jumbers to be ANDed are stored in the two registers of the circuit shown. Results are returned to he lower register.

8-1. Locate the letters a and b at the outputs of data latches A2 and B2. Use two sets of arrows to show the signal paths from a and b, through the ANDing process, to input D of data latch B2.

For lines 8-2 through 8-6, circle the correct answer.

8-2. T   F   AND ENABLE is active-high.

8-3. T   F   CLOCK must go active to store the results of the ANDing in Register B.

8-4. T   F   The flip-flops in Register A must be master-slave.

8-5. T   F   The flip-flops in Register B must be master-slave.

8-6. T   F   To load new numbers into the register, CLOCK must be activated.

8-7. If AND ENABLE and LOAD ENABLE were to be active at the same time, but CLOCK was inactiv
what would happen to the numbers in the registers?
   a. They would remain unchanged.
   b. The number in A would be ANDed with the one in B and the result stored in B.
   c. The number in A would be moved to B.

8-8. Show the action of the circuit by ANDing   A2A1A0   = 110
     the numbers at the right.   B2B1B0   = <u>010</u>

                                                       A2A1A0 AND B2B1B0 =

hapter 13
RITHMETIC/LOGIC CIRCUITS

Name _____

Date _____ Score _____

Instructor _____

FLAGS

-1. Indicate the parity of each of these numbers by circling even or odd.
   a. 1011 even/odd.
   b. 110,011 even/odd.
   c. 00,011,100 even/odd.

-2. ASCII codes are shown. They are seven bits long. When they are transmitted or stored, an eighth bit (a parity bit) is often added. Add parity bits to the following codes so even parity will result.

   a. ____100,0010.

   b. ____100,0101.

   c. ____010,1011.

-3. Which of the following is probably the most important use of parity?
   a. Decoding codes that contain large numbers of 1s.
   b. Helping determine the signs of numbers.
   c. Indicating errors in codes and numbers.

-4. The meaning of 1s in these four flags (parity, zero, carry, and sign) are shown at the right.

   If the number 1100,0011 were shifted one bit to the LEFT, what values (1s and 0s) would appear in these flags? The MSB will move into the carry flag. However, the carry flag is not part of the resulting number and does not enter into the determination of parity.

Flag	Implies
P = 1	Even parity
Z = 1	Results zero
C = 1	Carry out of MSB
S = 1	MSB equals 1

   P = __. Z = __. C = __. S = __.

-5. Repeat Problem 4 for this addition.

   P = __. Z = __. C = __. S = __.

   A = 10,010,000
   B = + 01,110,110
   Sum =

-6. Repeat Problem 4 for this addition.

   P = __. Z = __. C = __. S = __.

   A = 00,101,000
   B = + 11,011,000
   Sum =

-7. A flag register is shown at the right. Ones appear in the unused bits. If this register contained the hexadecimal number FA, what are the values of the four flags?

   P = __. Z = __. C = __. S = __.

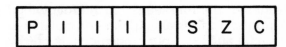

Chapter 13
ARITHMETIC/LOGIC CIRCUITS

Name _____

Date _____ Score _____

Instructor _____

PARITY (2-7486, 7404 or 2-4070, 4049)

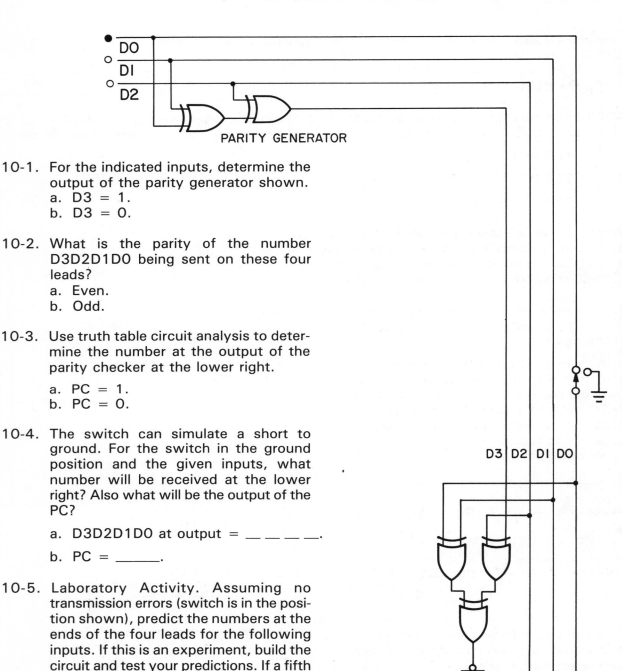

10-1. For the indicated inputs, determine the output of the parity generator shown.
  a. D3 = 1.
  b. D3 = 0.

10-2. What is the parity of the number D3D2D1D0 being sent on these four leads?
  a. Even.
  b. Odd.

10-3. Use truth table circuit analysis to determine the number at the output of the parity checker at the lower right.

  a. PC = 1.
  b. PC = 0.

10-4. The switch can simulate a short to ground. For the switch in the ground position and the given inputs, what number will be received at the lower right? Also what will be the output of the PC?

  a. D3D2D1D0 at output = __ __ __ __.

  b. PC = _____.

10-5. Laboratory Activity. Assuming no transmission errors (switch is in the position shown), predict the numbers at the ends of the four leads for the following inputs. If this is an experiment, build the circuit and test your predictions. If a fifth lamp is available, use it to monitor PC.

Input	Predic. output	Meas. output
D2D1D0	D3D2D1D0	D3D2D1D0
1 1 0	__ __ __ __	__ __ __ __
1 0 0	__ __ __ __	__ __ __ __

Chapter 14

BUSES

Name _____

Date _____ Score _____

Instructor _____

WIRED AND (7404, VOLTMETER)

1. The NOTs shown below drive a wired AND. Use arrows to show the flow of current from Vcc, through the common lead, to ground for each input signal set. If no current flows, skip that drawing.

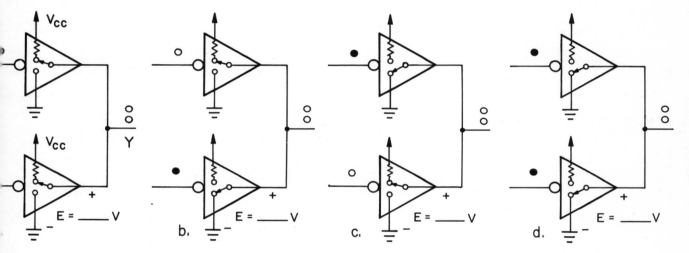

2. Laboratory Activity. For each input set in Problem 1, indicate the signal that will appear at the circuit's output. Use the upper dot to record each prediction. If this is an experiment, build the circuit and test your predictions. Use the lower dots to record your results.

3. Based on the values of Y from Problem 2, complete the truth table shown.

INPUTS		OUTPUT
A	B	Y
O	O	
O	I	
I	O	
I	I	

4. Which of the following elements is depicted by the truth table in Problem 3?
    a. AND.
    b. NAND.
    c. OR.
    d. NOR.

1-5. Add the symbol for a wired-AND to this diagram and write the logic expression for the complete circuit in terms of A, B, and Y. Then use DeMorgan's theorem to show that it is identical to the expression selected in Problem 4.

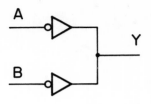

_____ = _____ = Y

1-6. Laboratory Activity. If this is an experiment, use a voltmeter to measure the signal voltag between the common lead and ground in the circuit of Problem 2. Record your results in t spaces provided in the drawing. If a signal is out of tolerance (logic 0 larger than 0.4 V; log 1 smaller than 2.4 V), circle that voltage.

Name _____

Date _____ Score _____

Instructor _____

OPEN-COLLECTOR ELEMENTS (7405, 7404, 1K, VOLTMETER)

2-1. Use arrows to show the several current paths between Vcc and ground in the circuit shown.

2-2. What is the critical disadvantage of the circuit in Problem 1?

   a. It operates too slowly.
   b. Its output can be 1 only when the outputs of all four elements are 1.
   c. The element outputting a 0 is likely to be overloaded.

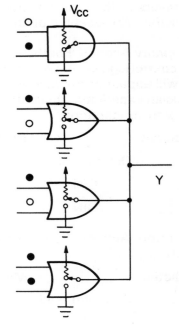

2-3. Laboratory Activity. For each input signal set, predict the signals on the common lead C. The driving elements are open-collector elements. Select the signals from the following set and record your predictions in the spaces marked A. If this is an experiment, build the circuit and test your predictions. Record your results in the spaces marked B.
   a. Logic 0 (less than 0.4 V).
   b. Floating lead (about 1.7 V).
   c. Logic 1 (more than 2.4 V).

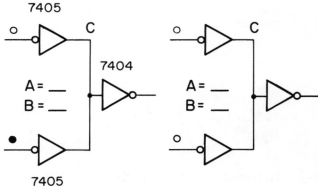

2-4. Laboratory Activity. Repeat Problem 3 for the circuit and signal sets shown.

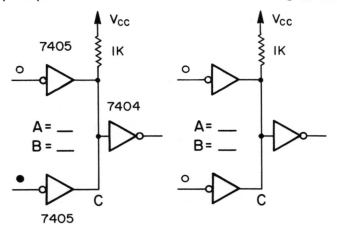

Chapter 14
BUSES

Name _____

Date _____ Score _____

Instructor _____

OPEN-COLLECTOR DRIVEN BUSES (7401, 7402, 1K)

The bus drivers in the circuit shown are open-collector NANDs. The gates at the receiving end a⟨
ordinary NORs. The enables at the receiving end are active-low.

3-1. Laboratory Activity. For the indicated data
and control signals, what numbers (1s and
0s) will appear at the following points? If
this is an experiment, build the circuit and
test your predictions.

Predicted		Measured
a. ____	Common lead	____ .
b. ____	CIN	____ .
c. ____	DIN	____ .

3-2. Laboratory Activity. Repeat Problem 1 for
AOUT = 0.

Predicted		Measured
a. ____	Common lead	____ .
b. ____	CIN	____ .
c. ____	DIN	____ .

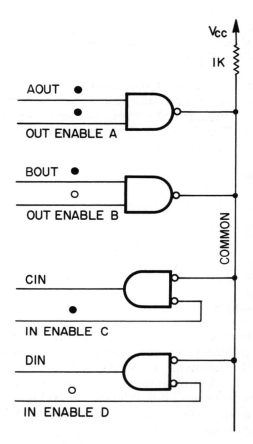

3-3. To transport a bit from BOUT to DIN, what numbers (1s and 0s) would be placed on the follow⟨
ing control leads?

a. ____ OUT ENABLE A.

b. ____ OUT ENABLE B.

c. ____ IN ENABLE C.

d. ____ IN ENABLE D.

3-4. Repeat Problem 3 for transporting a bit from AOUT to CIN.

    a. _____ OUT ENABLE A.

    b. _____ OUT ENABLE B.

    c. _____ IN ENABLE C.

    d. _____ IN ENABLE D.

3-5. Laboratory Activity. Assume the two drivers shown are the only elements driving the common lead. When OUT ENABLE A and OUT ENABLE B are both inactive, what is the state of the common lead? If this is an experiment, apply these control signals to your circuit and test your predictions.

    a. Logic 0.

    b. Logic 1.

    c. Floating lead.

**Chapter 14
BUSES**

Name _____

Date _____ Score _____

Instructor _____

THREE-STATE ELEMENTS

4-1. Match the following with the switch equivalents at the right.

    a. _____ Three-state AND.

    b. _____ Open-collector AND.

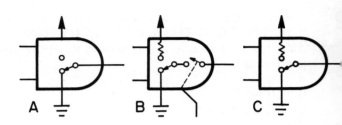

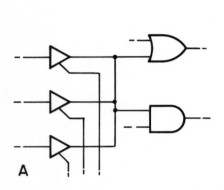 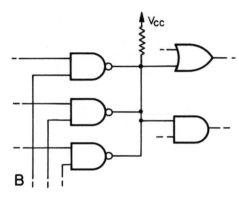 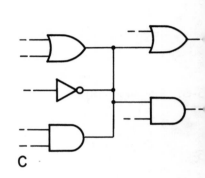

4-2. Match the following with the bus driving methods shown above.

    a. _____ Three-state.
    b. _____ Open-collector.

4-3. Use arrows to trace the paths of the currents that charge the stray capacitances in the circuit shown. There are two paths in each circuit. In each case, one path is from Vcc, through the internal circuit of the driven element, and out its input. The other paths are left to you. Assume the signal has just gone from 0 to 1.

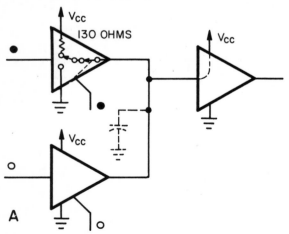 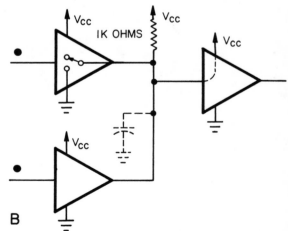

4-4. Based on the sizes of the resistors in Problem 3, which will be the faster circuit? That is, which circuit will have the higher charging current?
    a. A.
    b. B.
    c. They will be equal.

Name _____

Date _____ Score _____

Instructor _____

THREE-STATE BUS DRIVERS

A portion of a memory from a microcomputer is shown.

5-1. Which lead on a 6102 memory chip is probably its output?
    a. 11.
    b. 12.

5-2. For this memory to be read (output a number to the data bus), what signal must be on the clock lead?
    a. 1.
    b. 0.

5-3. Repeat Problem 2 for the starred lead.
    a. 1.
    b. 0.

5-4. The starred lead is attached to the computer's read/write lead. Which designation would be used on this lead?
    a. R/W.
    b. W/R.

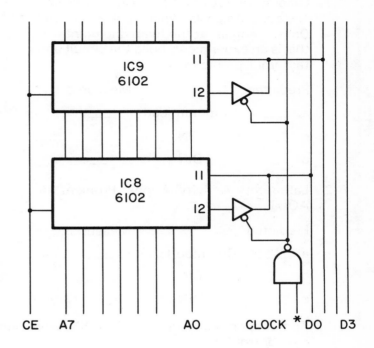

5-5. A block diagram of the above memory is shown. Add numbers near the slash marks to indicate the number of leads in each bus.

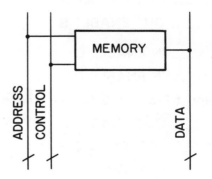

5-6. Which memory most likely has a built-in three-state driver at Dout?
    a. A.
    b. B.

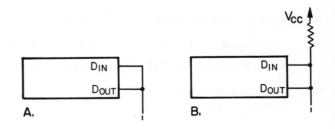

Chapter 14
BUSES

Name _____

Date _____ Score _____

Instructor _____

THREE-STATE BUS DRIVERS (7408, 74126 or 4081, 4503, VOLTMETER

The bus drivers in the circuit shown are three-state devices. Their output enables are active-high

6-1. Laboratory Activity. For the indicated data and control signals, what numbers (1s and 0s) will appear at the following points? If this is an experiment, build the circuit and test your predictions.

Predicted		Measured
a. _____	Common lead	_____.
b. _____	CIN	_____.
c. _____	DIN	_____.

6-2. Laboratory Activity. Repeat Problem for AOUT = 0.

Predicted		Measured
a. _____	Common lead	_____.
b. _____	CIN	_____.
c. _____	DIN	_____.

6-3. To transport a bit from BOUT to DIN, what numbers (1s and 0s) would be placed on the follow ing control leads?

a. _____ OUT ENABLE A.

b. _____ OUT ENABLE B.

c. _____ IN ENABLE C.

d. _____ IN ENABLE D.

6-4. Repeat Problem 3 for the transportation of a bit from AOUT to DIN.

a. _____ OUT ENABLE A.

b. _____ OUT ENABLE B.

c. _____ IN ENABLE C.

d. _____ IN ENABLE D.

6-5. Laboratory Activity. Assume that the two drivers shown are the only elements driving the com mon lead. When OUT ENABLE A and OUT ENABLE B are both inactive, what is the state o the common lead? If this is an experiment, apply these control signals to your circuit and tes your predictions.

a. Logic 0.

b. Logic 1.

c. Floating lead.

Chapter 14
BUSES

Name _____

Date _____ Score _____

Instructor _____

THREE-STATE BUS DRIVERS

The circuit shown represents two ports from a microcomputer. The computer is at the left; the outside world is at the right.

7-1. Which port is an output port?
   a. Port A.
   b. Port B.
   c. They are both output ports.

7-2. To activate Port A, what signals must be on the control leads?

   a. ____ ENABLE 1.

   b. ____ CLOCK.

   c. ____ ENABLE 2.

7-3. Why are the elements at Port A three-state while those at Port B are not?
   a. Those at A drive the bus; those at B take data from the bus.
   b. The elements at B are also three-state. However, there is no standard symbol for three-state flip-flops.
   c. As long as half the elements are three-state, the requirements of the bus have been met.

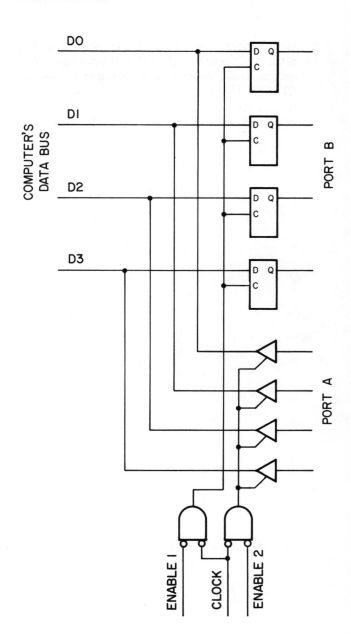

7-4. The bi-directional buffer/driver shown has three modes: transmit left to right, transmit right to left, and open circuit. For the control signals shown, in which mode is the circuit?
  a. L to R.
  b. R to L.
  c. Open.

7-5. One lead of the buffer/driver shown is an enable. The other controls the direction of signal flow. Which lead would be labeled DIRECTION?
  a. Lead A.
  b. Lead B.

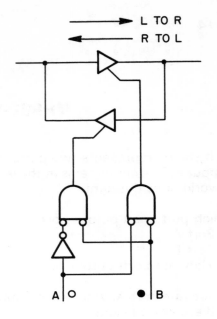

Name _____

Date _____ Score _____

Instructor _____

THREE-STATE BUS DRIVERS

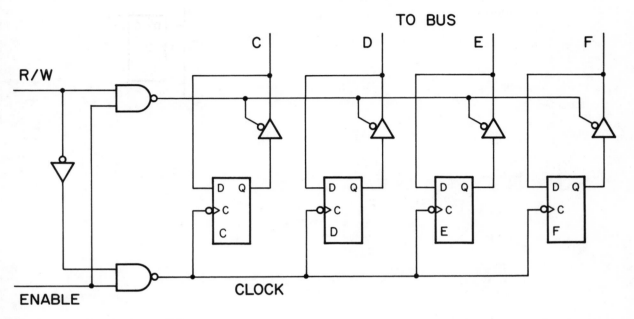

TO BUS

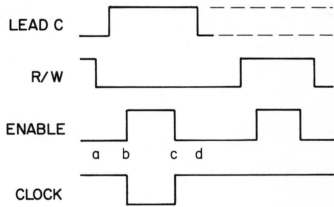

8-1. The register that is shown is connected directly to the bus. The timing diagram for data latch C is at the right. During the write portion of its cycle, a 1 is stored. At what point in the cycle does that storage take place?

a. Time a.
b. Time b.
c. Time c.
d. Time d.

8-2. During the read portion of the register's cycle, the 1 stored in Problem 1 is placed on the bus. Complete the graph of lead C to show the outputting of this 1. Darken in the graph for the times when the 1 is on the bus. Leave the dashed lines to show the times when the bus floats.

For True or False questions, circle the correct answer.

8-3. T   F   The above register can be read and written into at the same time.

8-4. T   F   If master-slave flip-flops were used, the connection at the right could be used. That is, the three-state element would not be needed.

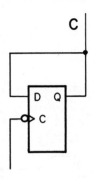

Name _____

Date _____ Score _____

Instructor _____

FANOUT (2-4704, VOLTMETER)

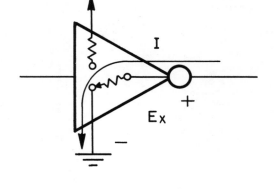

-1. Laboratory Activity. When the TTL NOT that is shown outputs a 0, it must sink current from the elements it drives. Because the resistance of the switch is small, Ex (the external signal voltage) is also small. However, the switch has some resistance. If this NOT is required to drive additional inputs (that is, I increases), what happens to Ex?
   a. Ex increases.
   b. Ex decreases.

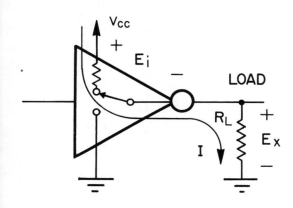

-2. Laboratory Activity. Refer to the above circuit. In it, RL represents the load placed on the NOT by the inputs it drives when outputting a 1. The more inputs it drives, the lower the resistance of RL and the higher the current I.

As I increases, Ei (the voltage across the internal resistance) also increases. Based on the following equation, what happens to Ex as inputs are added?

$$Ex = Vcc - Ei$$

   a. Ex increases.
   b. Ex decreases.

9-3. Laboratory Activity. Use the circuit at the right to test your predictions. For outputs of both 0 and 1, measure Ex with no load. Then load its output by adding inputs in groups of three. Power must be applied to both the driving and driven elements. Record the voltage measurements in the spaces provided. The total load of 11 inputs will not damage the driving element.

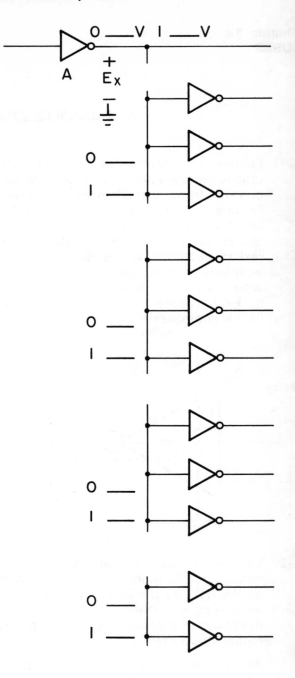

Chapter 14
USES

Name _____

Date _____ Score _____

Instructor _____

FANOUT AND FANIN

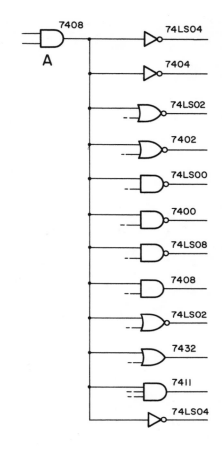

0-1. Refer to the circuit at the right. The fanout of a 7408 is 10. Does the load on the element at A exceed its rating?
a. Yes.
b. No.

0-2. Which of the following ANDs will be able to drive the larger number of inputs without exceeding a fanout of 10? Refer to the circuits marked A and B.
a. A (three-state output).
b. B (open-collector output).

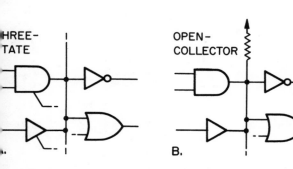

0-3. Indicate the fanin of each of the following leads of the shift register at the right.

a. _____ IN.

b. _____ CLEAR.

c. _____ SHIFT.

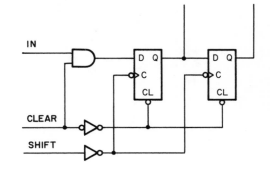

0-4. Where is fanin most likely to be a problem in the circuit shown?
a. At A (a point within the logic circuit).
b. At B (a point where a bus is to be driven).

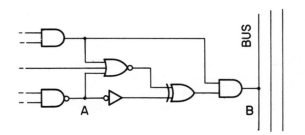

Chapter 15

READ-ONLY MEMORY

Name _____

Date _____ Score _____

Instructor _____

MEMORIES

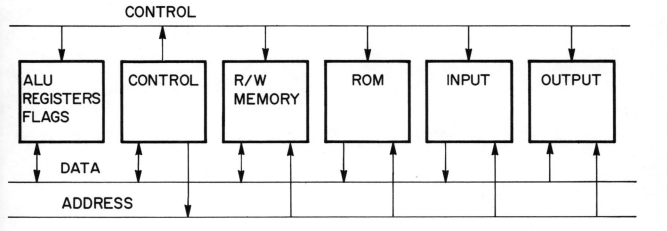

-1. The above block diagram represents a microprocessor-based computer. Where is the bootstrap program (the program used to control the loading of instructions and data) most likely stored?
   a. R/W memory.
   b. ROM.
   c. Control.

-2. In the above computer, which memory is most likely volatile?
   a. R/W memory.
   b. ROM.
   c. Both R/W and ROM are volatile.

-3. In the above drawing, only one arrow is shown on the lead between the data bus and the ROM. Why?
   a. It is a drafting error. There should be two as there are on R/W.
   b. ROMs are smaller than R/W memories, so they handle less data.
   c. Ordinarily, numbers from the data bus cannot be stored in ROM.

-4. From the standpoint of data flow, which input/output block might be treated as a read-only memory?
   a. Input.
   b. Output.

1-5. Based on the leads on the ROM shown, match the following functions with the indicated internal blocks.

a. _____ Address decoder.

b. _____ Output circuit.

c. _____ Memory array.

d. _____ Control circuit.

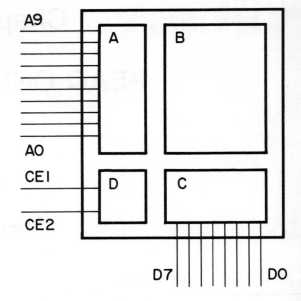

**Chapter 15
READ-ONLY MEMORY**

Name _____

Date _____ Score _____

Instructor _____

ESTIMATING MEMORY SIZE

-1. Estimate the size of each of the following memories. Express your answers in the form: number of words x bits per word.

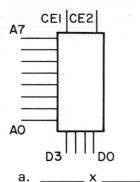
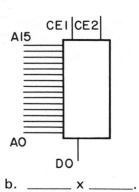
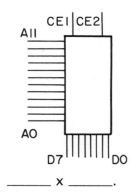

a. _____ x _____.   b. _____ x _____.   c. _____ x _____.

-2. Estimate the memory sizes of the following chips used in the following memories. Also, estimate the overall size of each memory.

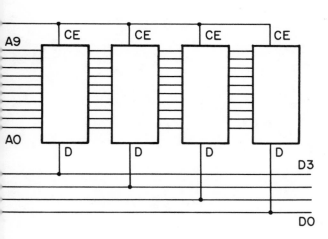
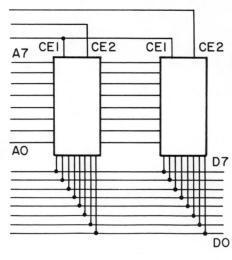

a. Chip size _____ x _____.   c. Chip size _____ x _____.

b. Memory size _____ x _____.   d. Memory size _____ x _____.

2-3. Repeat Problem 2 for the memory shown.

    a. Chip size   _____ x _____.

    b. Memory size   _____ x _____.

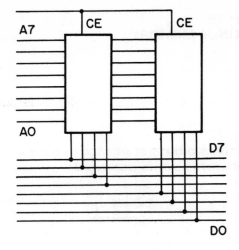

Chapter 15
READ-ONLY MEMORY

Name _____

Date _____ Score _____

Instructor _____

MEMORY ARRAYS

15-1. For the WORD ENABLE signals shown, determine the outputs of each of these memory arrays. Numbers in the boxes indicate the numbers stored in the cells.

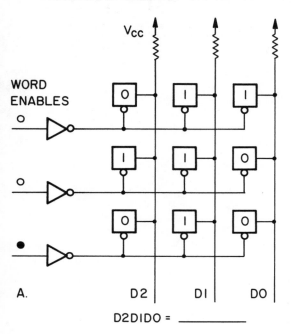

A.   D2   DI   DO

D2DIDO = _____

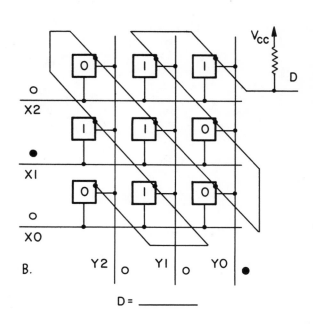

B.   Y2   o   YI   o   YO   ●

D = _____

15-2. Indicate the sizes of the above memories.

   a. A is _____ x _____.

   b. B is _____ x _____.

15-3. Which of the above memories is bit organized? (The other is word organized.)
   a. A.
   b. B.

15-4. Which most likely describes the bus drivers present in the cells of the above memory arrays?
   a. Three-state.
   b. Open-collector.

3-5. Indicate the overall size of the memory shown. Note that the X and Y leads are not address leads. They drive the X and Y leads of the matrices.

_____ x _____.

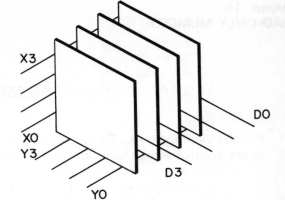

Chapter 15
READ-ONLY MEMORY

Name _____

Date _____ Score _____

Instructor _____

ARRAY ORGANIZATION TYPES (7404, 7408, 74126)

15-1. Laboratory Activity. Three-state buffers have been used as memory cells in the 2 x 2 ROM shown. Because there are only two words, only one address lead is needed.

For the control and address signals in the truth table shown, predict the outputs of the ROM. Use F to indicate floating output leads.

If this is an experiment, build the circuit and test your predictions. Depending on your trainer, floating leads may appear as 1s or 0s. You may have to use a logic probe to detect floating leads.

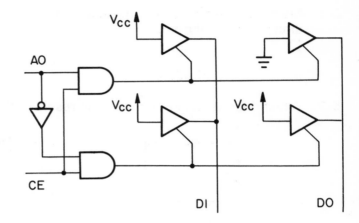

		PREDICTED		MEASURED	
AO	CE	DI	DO	DI	DO
I	I				
O	I				
I	O				
O	O				

15-2. Laboratory Activity. The array in the ROM shown is in the form of a matrix. Each cell consists of a three-state buffer and an AND. For the control signals shown in the table, predict the outputs of the ROM. Record the predictions in the chart provided.

If this is an experiment, build the circuit and test your predictions. Record the results in the chart provided.

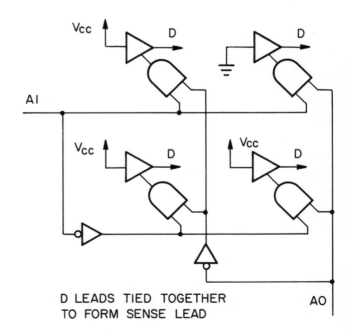

D LEADS TIED TOGETHER
TO FORM SENSE LEAD

ADDRESS		PREDICTED	MEASURED
AI	AO	D	D
O	O		
O	I		
I	O		
I	I		

Name _____

Date _____ Score _____

Instructor _____

DIODES AS MEMORY CELLS (7400, 5-1N914, 3-1K, VOLTMETER)

5-1. Match the following with the leads on the diode symbol shown.

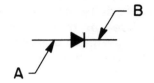

a. _____ Anode.

b. _____ Cathode.

5-2. Which bias (applied voltage) will result in the flow of current in the diode circuit shown?

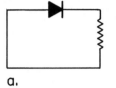 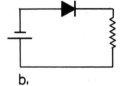 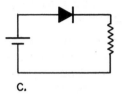

a.　　　　　　　b.　　　　　　　c.

5-3. Use a series of arrows to show the path (or paths) of current from Vcc to ground for the input signals shown at the right.

5-4. For the inputs shown, what number will appear at the output of the circuit from Problem 3?

D1D0 = _____.

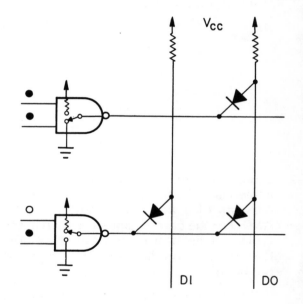

5-5. Laboratory Activity. For the input signals shown in the truth table, predict the outputs of the ROM shown. Record your predictions in the column marked Predic.

If this is an experiment, build the circuit and test your predictions. Also, with D2 = 0, measure the voltage between this output lead and ground. Is it less than the 0.4 volts required for a valid TTL logic 0?
a. Yes.
b. No.

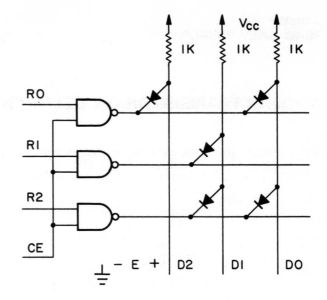

RO	RI	R2	CE	PREDIC. D2 DI DO	MEAS. D2 DI DO
I	O	O	I		
O	I	O	I		
O	O	I	I		
I	O	O	O		

Chapter 15
READ-ONLY MEMORY

Name _____

Date _____ Score _____

Instructor _____

BIPOLAR TRANSISTOR CELLS (7408, 4-2N2222, 4-2.2K, 3-1K, VOLTMETEI

6-1. Match the following with the leads of the transistor at the left below.

  a. _____ Emitter.　　b. _____ Base.　　c. _____ Collector.

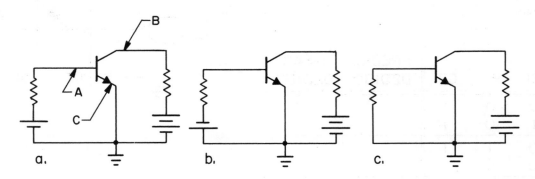

6-2. Refer to the circuit from Problem 1. Which base bias will result in current flow in the collect circuit?

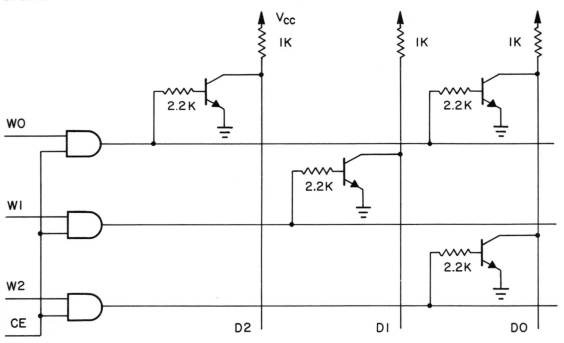

WO	WI	W2	CE	PREDIC. D2DIDO	MEAS. D2DIDO
I	O	O	I		
O	I	O	I		
O	O	I	I		
I	O	O	O		

6-3. Laboratory Activity. For the input signal sets in the truth table shown, predict the outputs of the above ROM.

If this is an experiment, build the circuit and test your predictions. Also, with DO = O, measure the signal voltage between this lead and ground. Is it less than the 0.4 volts required by a TTL logic O?
a. Yes.
b. No.

Name _____

Date _____ Score _____

Instructor _____

BIPOLAR TRANSISTOR MEMORY CELLS

7-1. Based on the action described in the following figure, which logic element best describes th
bipolar-transistor circuit shown?
a. Non-inverting buffer.
b. NOT.
c. OR.
d. AND.

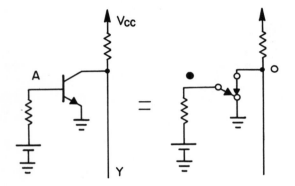

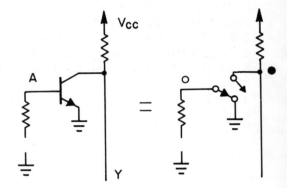

7-2. How are the problems associated with
wired-ANDs solved in the circuit shown?
a. The transistors act like open-collector
elements.
b. The transistors act like three-state
elements.

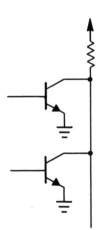

7-3. In the circuit shown, a three-state buf-
fer/driver (at C) is used between the common lead of the memory and the bus. For
each of the following input signal sets,
predict the signal that will appear on the
bus. Use F to indicate that the output of
the three-state element is open.

a. For A = 1, B = 0, OUT ENABLE = 1,

   number on bus = _____.

b. For A = 0, B = 1, OUT ENABLE = 1,

   number on bus = _____.

c. For A = 0, B = 0, OUT ENABLE = 0,

   number on bus = _____.

d. For A = 0, B = 0, OUT ENABLE = 1,

   number on bus = _____.

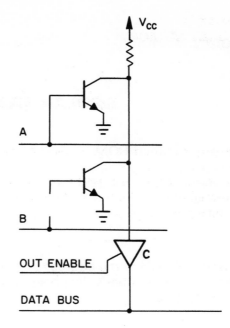

Chapter 15

READ-ONLY MEMORY

Name _____

Date _____ Score _____

Instructor _____

BIPOLAR TRANSISTOR MEMORY CELLS

A mask-programmed ROM is shown. The problems on this sheet refer to this circuit.

8-1. Determine the size of the memory. Give your answer in the form: number of words x bits per word.

_____ x _____.

8-2. For the signals at the address and control leads, what number will appear at the circuit's output?

$D1D0$ = _____.

8-3. Add connections to the bases of the bottom row of transistors to store the number $D1D0$ = 01.

8-4. Draw lines around the four sections of the memory (memory array, address decoder, output circuit, and control circuit). Label each.

8-5. If CE2 = 0 (assume A1 = 0, A0 = 1, CE1 = 1), what number would appear on lead D1?
   a. 1.
   b. 0.
   c. Floating lead.

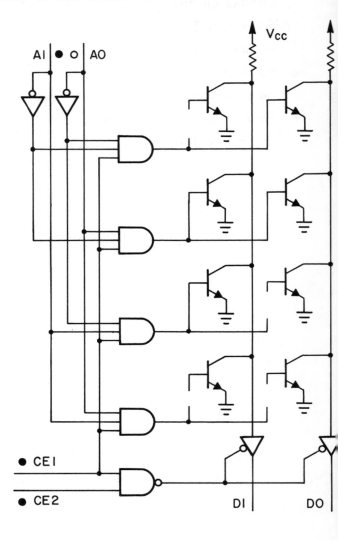

8-6. When CE1 = 0, what number appears on the common lead (on the array side of the three-state element) associated with D1?
   a. 1.
   b. 0.
   c. Floating lead.

The connection between CE1 and the ANDs in the address decoder is not necessary to the operation of this memory. No matter what address is applied, stored numbers cannot be delivered to the bus unless both CE1 and CE2 are active (high). However, when CE1 is inactive (low), the four ANDs output 0s and turn off all transistors. As a result, current flows to the memory array only when the circuit is being read. This reduces the average power consumed.

Chapter 15
READ-ONLY MEMORY

Name _____

Date _____ Score _____

Instructor _____

MOS TRANSISTOR MEMORY CELLS

9-1. Which circuit will output Y = 0? That is, in which circuit is the transistor conducting?
   a. A.
   b. B.
   c. In neither.

9-2. For the applied address and control signals, what number will appear at the output of the ROM shown?

   D2D1D0 = _____.

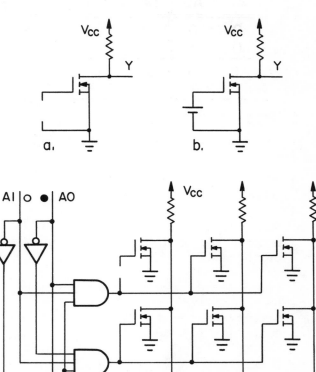

9-3. A MOS memory with an AO and enable decoder and a DATA OUT lead is shown. For each of the following addresses, determine the number on DATA OUT. Assume ENABLE = 1.

AI	AO	DATA OUT
O	O	
O	I	
I	O	
I	I	

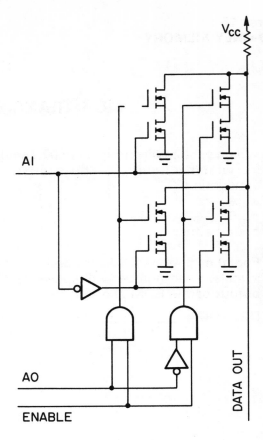

Chapter 15
READ-ONLY MEMORY

Name _____

Date _____ Score _____

Instructor _____

READ-ONLY MEMORIES

0-1. Match the following memories with the cells at the right.

a. _____ Mask-programmed diode ROM.

b. _____ Mask-programmed FET ROM.

c. _____ Bipolar PROM.

d. _____ EPROM.

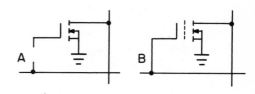

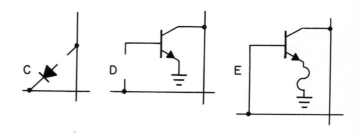

0-2. What is the probable purpose of the transparent window in the memory shown?
a. To permit inspection of the chip.
b. To allow the erasing of stored data.
c. To allow the pattern of 1s and 0s to be determined after programming.

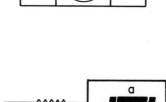

0-3. Memories can be used as encoders and decoders. The following diode array is used as an encoder. It drives the 7-segment display at the right. When an input lead (0 through 9) is energized, the proper segments are lit to display that decimal number. Add diode symbols to row 3 to light the proper segments to display a 3.

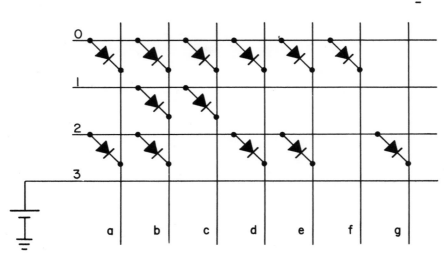

Chapter 16

RANDOM ACCESS MEMORY

Name _____

Date _____ Score _____

Instructor _____

MEMORIES BASED ON DATA LATCHES (7475, 74126, 7404, 2-7411)

oblems on this sheet refer to the small memory at the right.

1. Which of the following best describes the memory shown?
   a. ROM.
   b. RAM.

2. Which data latches are at address A = 1?
   a. 11,01.
   b. 01,00.
   c. 11,10.
   d. 10,00.

3. To store a number at A = 0, what numbers (1s and Os) must be applied to the control leads?

   a. CE = _____.

   b. W/R = _____.

4. Note the numbers at the outputs of the data latches (previously stored data). For the control signals indicated, what numbers will appear at the circuit's output? If these leads are floating, mark the outputs FL.

   a. DO1 = _____.

   b. DO0 = _____.

5. Note that the number 11 is at the input of this circuit. During the storage of this number, what numbers will appear at the circuit's output? If these leads are floating, mark the outputs FL.

   a. DO1 = _____.

   b. DO0 = _____.

6. How is the memory placed in its standby mode?
   a. A 0 is placed on CE.
   b. Zeros are applied to DI1 and DIO.
   c. W/R is allowed to float.

1-7. Why are data latches seldom used in large memories?
   a. They require too much space and power.
   b. They do not retain numbers as well as other memory elements.
   c. They are far too slow.

1-8. Laboratory Activity. If this is an experiment, build the circuit and test its operation. Store the numbers for part "a" below and then read them out. Then store the numbers for part "b" and read them out.

   a. Address   D1D0
      A = 1   00
      A = 0   11

   b. Address   D1D0
      A = 1   10
      A = 0   01

**Chapter 16
RANDOM ACCESS MEMORY**

Name _____

Date _____ Score _____

Instructor _____

BIPOLAR-TRANSISTOR MEMORY CELL (2-2N2222, 2-3.3K, 2-10K)

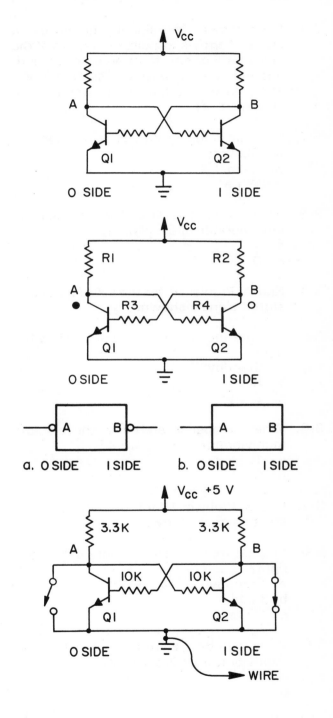

2-1. Assume Q1 is conducting (a 0 has been stored).
   a. Use a series of heavy arrows to show the path of the collector current from Vcc to ground.
   b. Use light arrows to show the path of the base current from Vcc to ground.

2-2. An active-low circuit is shown. If a 1 has been stored, a 0 will appear on the 1 side. If a 0 has been stored, the 0 will appear on the 0 side. Based on the indicated measurements, a 1 is stored in the cell. Which transistor is conducting?
   a. Q1.
   b. Q2.
   c. Both.

2-3. Which memory cell best represents the memory cell in Problem 2?

2-4. When switch B in the circuit shown is opened, which transistor will be conducting?
   a. Q1.
   b. Q2.
   c. Both.

2-5. Assume that the switch at A remains open. What would happen if the switch at B were opened and closed a number of times?
   a. The stored number would remain unchanged.
   b. The stored number would alternate between 1 and 0.

2-6. Laboratory Activity. If this is an experiment, build the circuit from Problem 4 and test its action. Touch the ground wire alternately to points A and B to store 0s and 1s.

Chapter 16
RANDOM ACCESS MEMORY

Name _____

Date _____ Score _____

Instructor _____

MULTIPLE-EMITTER MEMORY CELL

3-1. For the measured signals shown, use a series of arrows to indicate the path of the collector current from Vcc to ground. Ignore the base current. (Remember, a 1 at the base of a transistor turns it on; a 0 turns it off.)

3-2. In what mode is the cell at the right?
   a. Standby.
   b. Read.
   c. Write.

3-3. What number is stored in the bistable multivibrator of Problem 1?
   a. 1.
   b. 0.

3-4. Repeat Problem 1 for the set of input and stored signals shown.

3-5. In what mode is the memory cell at the right?
   a. Standby.
   b. Read.
   c. Write.

3-6. What number is stored in the bistable multivibrator of Problem 4?
   a. 1.
   b. 0.

3-7. Repeat Problem 1 for the set of input and stored signals shown.

3-8. In what mode is the memory cell at the right?
   a. Standby.
   b. Read.
   c. Write.

3-9. What number is stored in the bistable multivibrator of Problem 7?
   a. 1.
   b. 0.

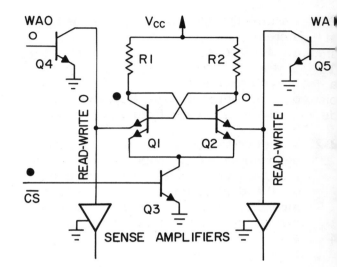

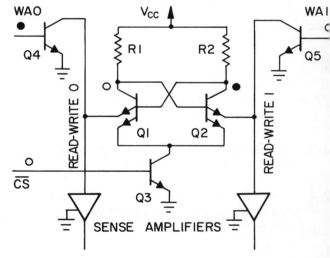

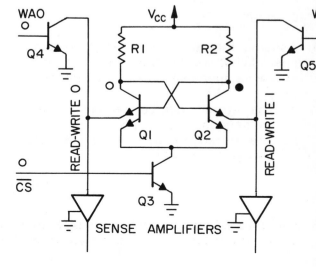

Chapter 16
RANDOM ACCESS MEMORY

Name _____

Date _____ Score _____

Instructor _____

MULTIPLE-EMITTER MEMORY ARRAY

Problems on this sheet refer to the memory array at the right. Its cells are identical to those on the sheet marked 3-1 through 3-9 (called Problem 16-3-1 through Problem 16-3-9). A letter C in a cell indicates a conducting transistor in the bistable multivibrator. An N indicates a nonconducting transistor. Cell outputs are active-low.

4-1. Word 1 is in its standby mode. That is, WS = 1 and the D inputs are all inactive. Use a series of arrows to show the paths of current from Vcc to ground through cells 11 and 10.

4-2. Word 2 is in its read mode. WS = 0 and the D inputs are all inactive. Use a series of arrows to show the paths of current from Vcc through cells 21 and 20 to ground.

4-3. What four bits must be applied to the data leads to store the number 10 in word 3?
  a. D10 = _____.
  b. D11 = _____.
  c. D00 = _____.
  d. D01 = _____.

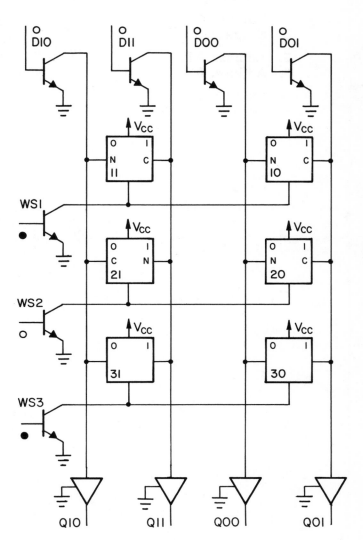

4-4. Address, control, and input/output circuits have been added to the above array to produce the memory at the right. Place the following letters on the leads of the memory to indicate their functions.

A. Input D1.
B. Input D0.

C. Output D1.
D. Output D0.

E. Address A1.
F. Address A2.

G. Chip enable CE.
H. Read/write R/W.

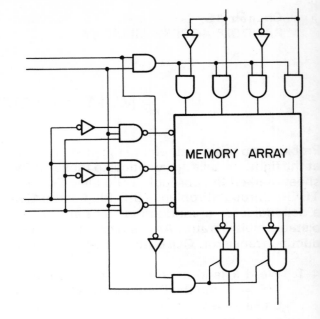

**Chapter 16
RANDOM ACCESS MEMORY**

Name _____

Date _____ Score _____

Instructor _____

MEMORY CELLS BASED ON FETs

16-1. Measured signals at the drains of Q1 and Q2 are shown at the right. Use a series of arrows to show the flow of drain current from Vdd to ground. (Because FETs are used, there is no gate current.)

16-2. Refer to the circuit from Problem 1. Which transistor has the higher (more positive) voltage applied to its gate?
a. Q1.
b. Q2.
c. Both gates are grounded.

16-3. In the circuit shown, transistors Q3 and Q4 have replaced resistors R1 and R2. For the measured signals, which transistors will be conducting? Mark as many as necessary.
a. Q1.
b. Q2.
c. Q3.
d. Q4.

16-4. In the circuit of Problem 3, why were transistors used in place of the resistors?
a. Transistors are active devices and add to the speed of the circuit.
b. Transistors require less space on an IC chip.
c. Transistors dissipate less heat than do resistors.

16-5. A 1 is being written into the cell shown. Circle all transistors that are conducting. See Fig. 16-12 of the textbook.

16-6. Refer to the circuit at the right. Use a series of arrows to show the path of current from Vdd to the ground at the source of Q8.

16-7. Again refer to the circuit at the right. When CELL ENABLE and WRITE ENABLE return to 0 (inactive), which transistors will continue to conduct? There will be more than one. See Fig. 16-12 of the textbook.
a. Q1.   e. Q5.
b. Q2.   f. Q6.
c. Q3.   g. Q7.
d. Q4.   h. Q8.

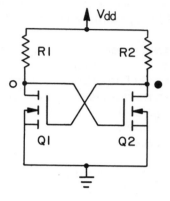

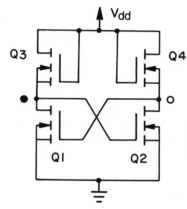

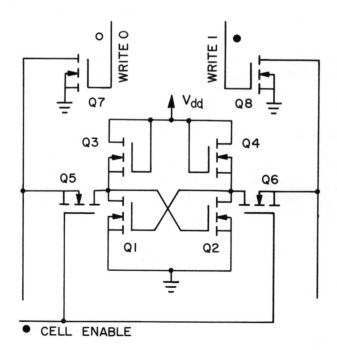

Chapter 16
RANDOM ACCESS MEMORY

Name _____

Date _____ Score _____

Instructor _____

MEMORY CELLS BASED ON CMOS FETs (4049)

6-1. Place an n and a p in the blanks under the MOS FET symbols shown.

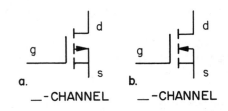

6-2. Which transistor in Problem 1 is turned on when its gate is positive with respect to its source? (The other transistor is turned on when its gate is negative with respect to its source.)
a. N-channel.
b. P-channel.

6-3. Refer to the diagrams marked A and B. Which gate voltage will turn the transistor on? Note that its source is connected to Vdd.
a. A.
b. B.

6-4. Refers to the diagrams marked "a." and "b" with Q1, Q2, Q3, and Q4. For each applied signal at the right, circle the transistors that will be turned on.

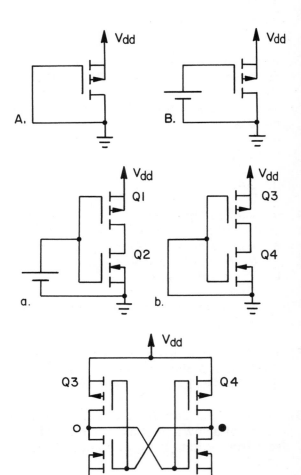

6-5. Refer to the diagram with Q3 above Q1 and Q4 above Q2. For the measured signals shown, circle the transistors that will be turned on.

6-6. Refer to the circuit at the right. Use a series of arrows to show the path of drain current from Vdd to ground. If there is no current (that is, there is no complete circuit), mark the drawing, "No drain current."

6-7. Laboratory Activity. The 4049 is a hex, CMOS NOT. A simplified circuit for one NOT is shown at the right. Two such NOTs can be used to build the circuit in Problem 5. Show how the two NOT symbols would be connected to produce a CMOS FET bistable multivibrator. If this is an experiment, build the circuit and test its action. Alternately ground pins 2 and 4 to input 0s and 1s.

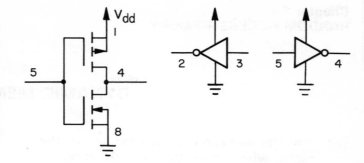

Chapter 16
RANDOM ACCESS MEMORY

Name _____

Date _____ Score _____

Instructor _____

DYNAMIC MEMORY CELL

7-1. For the measured signals shown at the
right, which transistor is conducting?
   a. Q1.
   b. Q2.

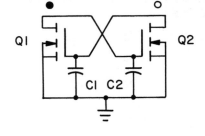

7-2. In the circuit at the right, which capacitor
is charged?
   a. C1.
   b. C2.

7-3. How are the capacitors in the above cell placed on the IC chip?
   a. They are part of the FET.
   b. They are manufactured separately and attached during final assembly.
   c. They are so large they cannot be placed on the chip. They are placed on the printed circu
   board beside the chip.

7-4. The circuit shown is in its write mode.
CELL ENABLE and W1 (write 1) are active,
so a 1 is being stored. The arrows show
the charging of C2. Use another series of
arrows to show the discharging of C1.
Start at C1 and proceed to ground.

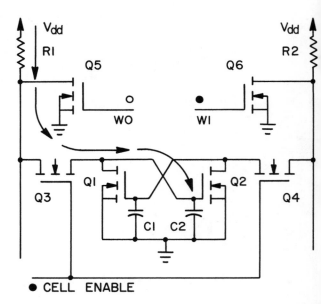

-5. The displayed cell is in its read mode, since CELL ENABLE is active and W1 is inactive. W0 is inactive. See part "c" of Fig. 16-20 from the textbook. Due to the charge on C2, Q2 is conducting. Refer to the arrows. As a result, READ-WRITE 1 is active (low).

When this cell is read, it is also refreshed. Use a series of arrows to show the charging current flowing from Vcc to C2.

-6. What happens if a dynamic cell is not refreshed?
   a. The stored bit is lost.
   b. The cell will be damaged.
   c. Power must be turned off and then restored before the stored data can be read.

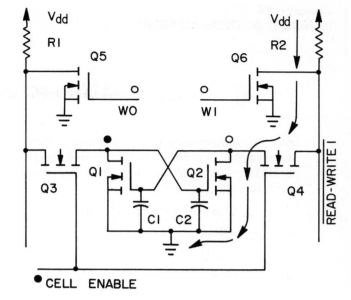

Chapter 16
RANDOM ACCESS MEMORY

Name _____

Date _____ Score _____

Instructor _____

RAM AND ROM MEMORY CELLS

8-1. Match the following with the memory cells at the right. A given cell may be used more than once. If there is more than one correct answer, give only one.

a. ____ Mask-programmed ROM.

b. ____ PROM.

c. ____ CMOS FET RAM.

d. ____ EPROM.

e. ____ Dynamic RAM.

f. ____ Bipolar technology.

g. ____ Uses transistors as resistors.

h. ____ Very low standby power.

i. ____ Ultraviolet light used to remove stored numbers.

j. ____ Must be refreshed.

k. ____ Programmed by manufacturer.

l. ____ Nonvolatile.

8-2. It takes time to read a stored number. Which of the following circuits would be expected to take the most time to deliver an 8-bit word to a bus?
a. Register.
b. RAM memory.

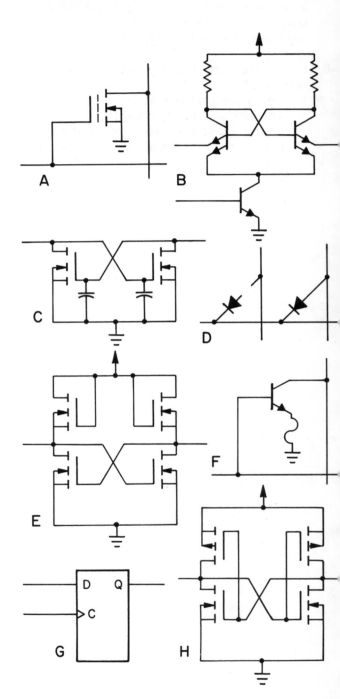

3-3. Which pair of leads is most likely the input/output of the circuit shown?
   a. A-A.
   b. B-B.
   c. C-C.

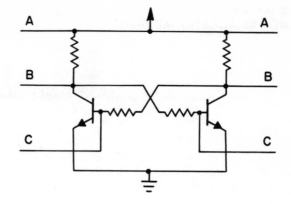

Chapter 17

MAGNETIC MEMORY

Name _____

Date _____ Score _____

Instructor _____

ELECTROMAGNETIC FIELDS

1-1. The arrows indicate the flow of current in the three legs of the circuit shown. Which point in the circuit will probably display the stronger magnetic field?
a. A.
b. B.
c. C.

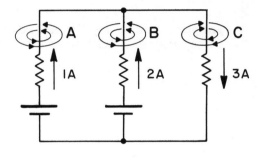

1-2. Repeat Problem 1 for the circuit with a 4-loop coil, a straight wire at A, a battery, and a resistor.
a. A.
b. B.

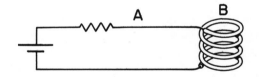

1-3. Repeat Problem 1 for the circuit with a 3-loop coil, a 7-loop coil, a resistor, and a battery.
a. A.
b. B.

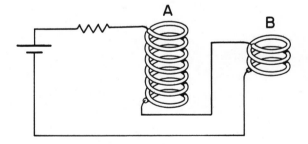

1-4. Repeat Problem 1 for the circuit with a 4-loop coil in air, a 4-loop coil around (linking) an iron core, a resistor, and a battery.
a. A.
b. B.

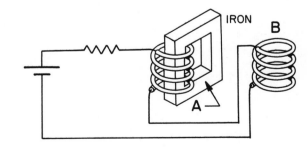

1-5. Repeat Problem 1 for the circuit with a paired center wire arrangement, two batteries, and two resistors.

 a. A.

 b. B.

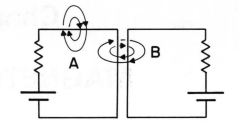

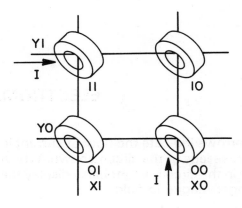

1-6. Four wires pass through the four small iron toroids shown. A torus has a doughnut shape. Current is flowing in wires X0 and Y1. In which toroid is the magnetic field likely to be the strongest?

 a. 00.

 b. 01.

 c. 10.

 d. 11.

 e. Toroids 11, 10, and 00 will have the same magnetic field strengths.

Chapter 17
MAGNETIC MEMORY

Name _____

Date _____ Score _____

Instructor _____

MAGNETIZATION CURVE

2-1. The magnetization curve for a sample of iron is shown. Match the following with the lettered points on the curve.

    a. ____ Axis on which NI (ampere-turns) is plotted.

    b. ____ Axis on which total magnetism is plotted.

    c. ____ Saturation.

    d. ____ Residual magnetism.

2-2. Draw a dashed line on the graph of Problem 1 to represent the magnetization curve of air.

2-3. Assume that point A represents the magnetization in the toroid shown. Use a series of arrows to show the path that would be followed to bring the magnetization to B. Is the path on the magnetization curve clockwise or counterclockwise?

2-4. Which of the following terms is used to describe the path indicated in Problem 3?
    a. Residual magnetism.
    b. Saturation.
    c. Hysteresis.

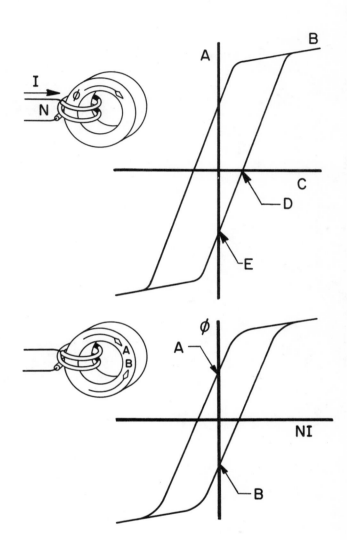

2-5. Which curve displays the larger residual
magnetism?
a. A.
b. B.
c. They are equal.

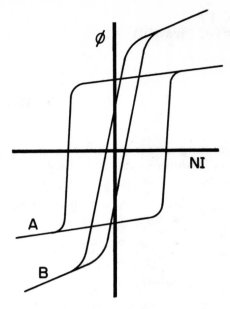

Chapter 17
MAGNETIC MEMORY

Name _____

Date _____ Score _____

Instructor _____

INDUCED VOLTAGE

3-1. The following shows three wires moving in magnetic fields. Which wire will cut the largest number of magnetic lines? Wire motion is indicated by the large arrows.
    a. A.
    b. B.
    c. C.
    d. They are all the same.

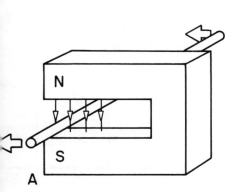
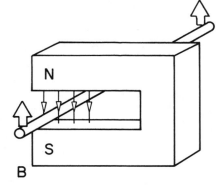
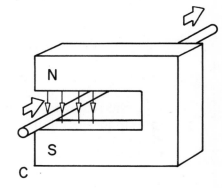

3-2. Which wire in Problem 1 will have the highest induced voltage between its ends?
    a. A.
    b. B.
    c. C.
    d. The voltages will be the same.

3-3. The wire at A is moving to the left. At B the magnet is moving to the right. Their speeds are equal. Which wire will have the higher voltage induced between its ends?
    a. A.
    b. B.
    c. They will be equal.

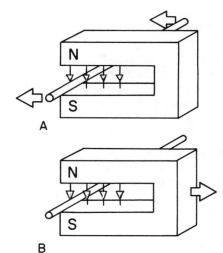

3-4. The displayed illustration depicts the magnets in the film on magnetic recording tape. The large arrow indicates that the tape is moving to the right. The fine lines represent the magnetic fields near the tape. The black dots are end views of stationary wires. For the instant shown, which wire will have the higher induced voltage? That is, which wire is cutting the most magnetic lines?
    a. A.
    b. B.

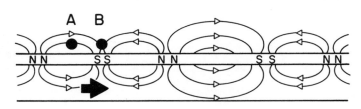

Chapter 17
MAGNETIC MEMORY

Name _____

Date _____ Score _____

Instructor _____

INDUCED VOLTAGE AND MAGNETISM FUNDAMENTALS

4-1. Which of the following actions will NOT induce a voltage between the ends of the wire at the left of each pair?

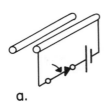

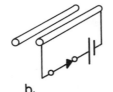

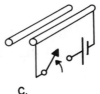

a. CLOSING SWITCH   b. HOLDING SWITCH CLOSED   c. OPENING SWITCH

4-2. The magnetization in the iron toroid shown is to be shifted to point C. Which transition will result in the greater change in total magnetization?
   a. From A to C.
   b. From B to C.

4-3. Which transition from Problem 2 will result in the greater induced voltage in the straight wire?
   a. From A to C.
   b. From B to C.

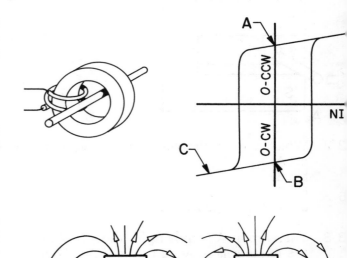

4-4. Lines of magnetic force are assumed to come out of the north end of a magnet and return to its south end. See the first drawing at the right. If a piece of iron is placed in a magnetic field, magnetic lines tend to pass through the iron rather than the air. See the second drawing. This causes the iron to act like a magnet. Which end of the piece of iron will be its north pole?
   a. A.
   b. B.

4-5. A portion of a recording head positioned above a magnetic tape is shown. The magnetic field around the head extends into the magnetic film on the tape. As a result, a small portion of the film becomes magnetized. Which end of the resulting small magnet will be its north pole?
   a. A.
   b. B.

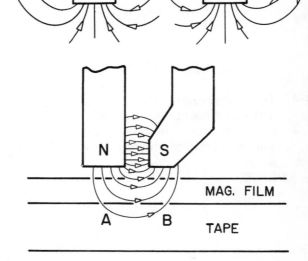

Chapter 17
MAGNETIC MEMORY

Name _____

Date _____ Score _____

Instructor _____

FLOPPY DISK

5-1. Match the following functions with the features of the floppy disk shown.

a. _____ Drive-spindle hole.

b. _____ Head access.

c. _____ Index hole.

d. _____ Disk.

e. _____ Jacket.

f. _____ Write-protect.

g. _____ Liner.

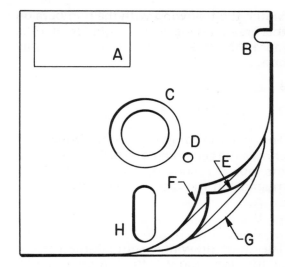

5-2. Which describes the relationship between the disk and jacket once inserted into a disk drive?
a. The disk rotates within the jacket.
b. The drive mechanically removes the disk from the jacket.
c. The operator removes the jacket before inserting the disk.

5-3. What produces the concentric lines that can be seen on some disks that contain stored data?
a. These are scratches caused by the head rubbing on the disk.
b. These are scratches caused by the liner rubbing on the disk (especially if the disk is dirty).
c. If overdriven, the magnetic tracks become visible.
d. These are the small grooves in which data is stored.

5-4. Mark each of the following statements about the care of floppy disks true or false. Circle the correct answer.

a. T F It is all right to touch disks through the head-access slot if one's hands are clean.

b. T F The jacket will protect the disk from all but the strongest magnetic fields.

c. T F The permissible temperature range for disks in storage is 0 °F to 212 °F.

d. T F Because disks are floppy, they can be bent rather sharply without damaging the film.

e. T F When not in use, disks should be kept on a flat surface or left in a disk drive.

f. T F Dust is not a problem if care is taken to dust disks with a tissue each time they are placed in a drive.

**Chapter 17
MAGNETIC MEMORY**

Name _____

Date _____ Score _____

Instructor _____

FLOPPY DISK STORAGE

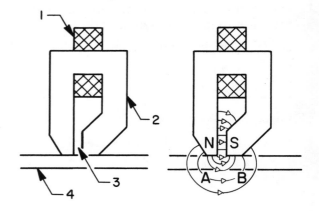

6-1. Match the following with the numbers on the first drawing at the right. If there is more than one correct answer, give only one.

    a. _____ Coil.

    b. _____ Gap.

6-2. The second drawing from Problem 1 shows the recording of a bit on a disk. For the polarity shown, which end of the magnet on the disk will be north?
    a. A.
    b. B.

6-3. The three drawings with six read heads at the right show a track moving under two read heads. For the head positions shown, which will have the larger induced voltage?
    a. A.
    b. B.

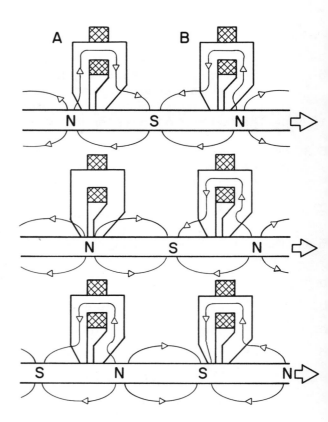

6-4. Which of the following best describes a stepping motor?
    a. The first pulse starts such motors. They stop on the second.
    b. Each time a stepping motor is pulsed, it turns a fixed number of degrees.
    c. They turn at fixed speeds of 3600, 1800, and 1200 rpm.

6-5. Which of the following is usually the smallest amount of data that can be read from or written to a floppy disk at one time?
    a. Bit.
    b. Byte.
    c. Sector.
    d. Track.

6-6. Which disk is probably hard sectored?
    a. A.
    b. B.

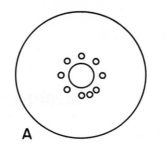
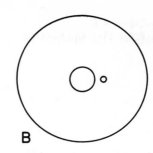

A   B

6-7. T   F   Formatting a disk is similar to refreshing a dynamic memory. Both must be accomplished at intervals or stored data will be lost. Circle the correct answer.

Chapter 17
MAGNETIC MEMORY

Name _____

Date _____ Score _____

Instructor _____

BUBBLE MEMORY

7-1. Which best describes a bubble memory?
   a. Magnetic random-access memory (individual words can be addressed and directly read fro⟨
      memory).
   b. Serial magnetic memory (stored data is moved to read heads before desired data is read⟩

7-2. Which magnetic memory is likely to have the least mechanical problems?
   a. Bubble.
   b. Disk.
   c. Tape.

7-3. Match the following parts of a bubble memory with the drawing shown. If there is more than one correct answer, give only one.

   a. ____ Active film (where bubbles are).

   b. ____ Substrate.

   c. ____ Permanent magnet.

   d. ____ Electromagnet.

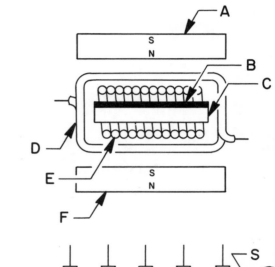

7-4. The displayed drawing shows the active film of a bubble memory and the applied magnetic field. Which darkened area best represents a domain (bubble used to store a 1)?
   a. A.
   b. B.
   c. C.

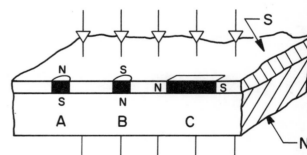

7-5. Which of the following is most likely used as an active film in the bubble memory in Problem 4?
   a. Silicon.
   b. Germanium.
   c. Garnet.
   d. Aluminum.

6. The displayed drawing shows one type of bubble generator. When it is pulsed, a bubble is formed (a 1 is stored). If it is not pulsed, no bubble is formed, and a 0 is stored. Which of the following is probably applied to form a bubble?
   a. Voltage.
   b. Current.

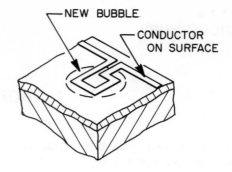

NEW BUBBLE

CONDUCTOR ON SURFACE

Chapter 17
MAGNETIC MEMORY

Name _____

Date _____ Score _____

Instructor _____

BUBBLE MEMORY

8-1. A piece of iron has been placed in the magnetic field at the right. As a result, it acts like a small magnet. Fill in the proper dot to indicate its north pole.

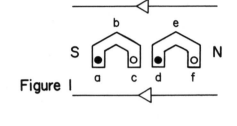

● NORTH POLE
○ SOUTH POLE

8-2. Chevrons, similar to those at the right, are used in some bubble memories to move magnetic bubbles. Like the piece of iron in Problem 1, these chevrons become magnets when placed in an external magnetic field. The dots indicate the polarities of positions a through f.

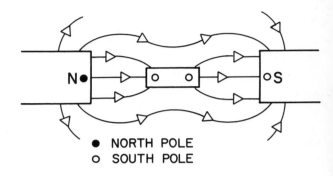

Figure I

For each of the following external magnetic fields, use open and solid dots like those in Figure 1 to indicate the polarities of the positions on the chevrons shown. Start at the position just above ''a'' and draw dots for side legs and pointed tops.

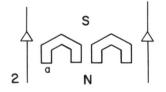

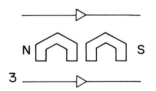

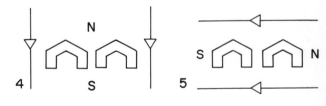

8-3. Assume the bubble at the right (with SOUTH at the top and NORTH at the bottom in the ''120°/120°/120°'' isometric type of drawing shown) is under point a in Figure 1 in Problem 2 (called Problem 17-8-2). Use a series of circles to show the positions of this bubble in Figures 2 through 5.

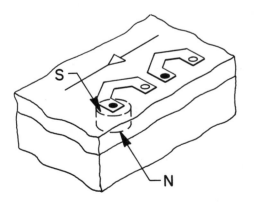

-4. In which of the following patterns would
you expect the chevrons to be arranged
in a present-day bubble memory?
    a. A.
    b. B.
    c. C.

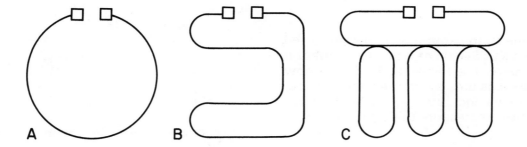

Chapter 17
MAGNETIC MEMORY

Name _____

Date _____ Score _____

Instructor _____

STORING DATA IN CORE MEMORY

9-1. At time 0, the stored bit in the displayed core is a 0. See point A on the magnetization curve. The following drawings show this core as time progresses. For each set of currents, indicate the point on the curve that best represents the condition of the core.

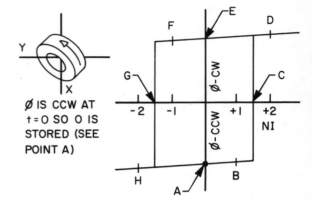

Ø IS CCW AT t=0 SO 0 IS STORED (SEE POINT A)

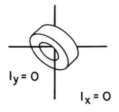

TIME 1 POINT ___

$I_y = 0$

$I_x = +1$ UNIT OF CURRENT

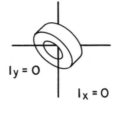

TIME 2 POINT ___

$I_y = 0$

$I_x = 0$

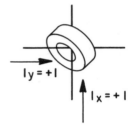

TIME 3 POINT ___

$I_y = +1$

$I_x = +1$

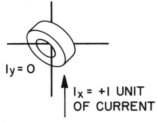

TIME 4 POINT ___

$I_y = 0$

$I_x = 0$

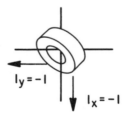

TIME 5 POINT ___

$I_y = -1$

$I_x = -1$

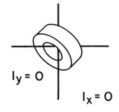

TIME 6 POINT ___

$I_y = 0$

$I_x = 0$

9-2. The arrows at the right indicate the directions of + current flow (the direction that results in the storing of a 1). Indicate the leads that must be energized to store a 1 in the core in the square. Also indicate the polarity of the current. Circle + or −.

a. Lead X ____ +/−.

b. Lead Y ____ +/−.

9-3. Repeat Problem 2 for the storing of a 0 in the circled core.

a. Lead X ____ +/−.

b. Lead Y ____ +/−.

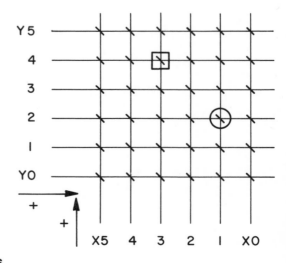

Chapter 17
MAGNETIC MEMORY

Name _____

Date _____ Score _____

Instructor _____

READING DATA FROM CORE MEMORY

10-1. To describe core memory, circle one word in each of the following pairs.

Select one:

a. Volatile.

b. Nonvolatile.

Select one:

c. Destructive read.

d. Nondestructive read.

Select one:

e. Random access.

f. Serial access.

10-2. To read the displayed core, negative unit currents are applied to both X and Y. Which stored number, 1 or 0, will result in the greater change in magnetism and thus the greater induced voltage in the sense wire?

a. Stored 1.

b. Stored 0.

c. They are equal.

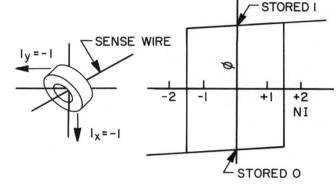

10-3. After being read, what number will appear in the above core?

a. The number that was there before it was read.

b. The complement of the number that was there before it was read.

c. 1.

d. 0.

10-4. The arrows show the direction of + current (the current that will store a 1). Indicate the leads that must be energized and the direction of the currents to read the circled core.

a. Lead X _____ +/−.

b. Lead Y _____ +/−.

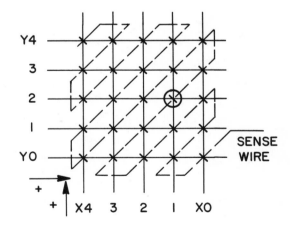

10-5. What is the size of the core memory shown? Give your answer in the form: number of words x bits per word.

_____ x _____

10-6. Which set of leads is used during the store process?
  a. Inhibit wires.
  b. Sense wires.

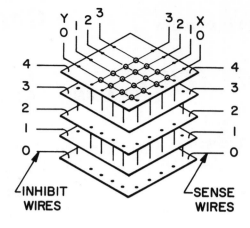

Chapter 18

INPUT/OUTPUT FOR SMALL COMPUTERS

Name _____

Date _____ Score _____

Instructor _____

PORT-ORIENTED INPUT/OUTPUT

The displayed circuit is an input and an output from a small computer.

1-1. On which side of this circuit is the computer most likely located?
  a. A.
  b. B.
  c. On either side.

1-2. Which control lead is likely to be marked IN? (The other will probably be marked OUT.)
  a. C.
  b. D.

1-3. What is the port number of the input port? Give your answer in both binary and hexadecimal.

A7 A6 A5 A4 A3 A2 A1 A0   Hex.

— — — — — — — — = —

1-4. Repeat Problem 3 for the output port.

A7 A6 A5 A4 A3 A2 A1 A0   Hex.

— — — — — — — — = —

1-5. For the control, address, and data signals shown, what number will appear on the data bus?

D3 D2 D1 D0

— — — —

1-6. The above circuit has eight address leads. Based on these leads, how many input ports could be connected to this computer?

Maximum number of input ports = ____.

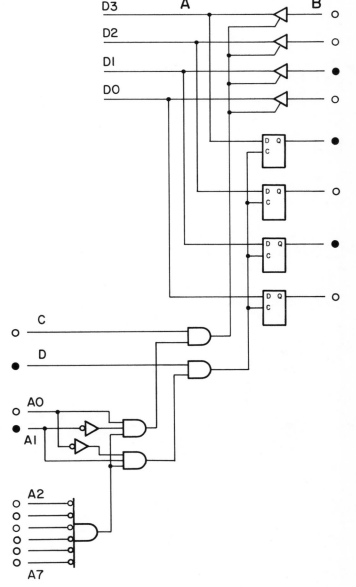

1-7. The decoder from the circuit on page 239 has been reproduced at the right. For the input signals shown, which PORT SELECT lead will be active?

    a. E.

    b. F.

    c. Neither.

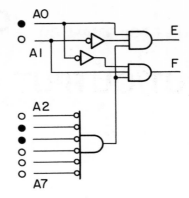

Chapter 18
INPUT/OUTPUT FOR SMALL COMPUTERS

Name _____

Date _____ Score _____

Instructor _____

MEMORY-MAPPED INPUT/OUTPUT

The displayed circuit represents two memory-mapped inputs.

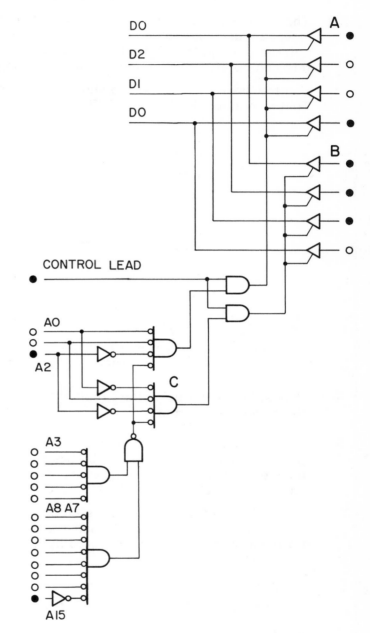

2-1. Which of the following instructions would NOT normally be used to input data through the circuit at the right?
   a. IN   (input).
   b. LDA   (load accumulator from memory location).
   c. ADM (add number from memory to number in accumulator).

2-2. To which computer lead would the control lead of the circuit most likely go?
   a. IN.
   b. MEMORY READ.
   c. OUT.
   d. MEMORY WRITE.

2-3. What are the addresses of these inputs? Give your answers in hexadecimal.

   Address of A = __ __ __ __.

   Address of B = __ __ __ __.

2-4. For the signals shown, what number will appear on the circuit's data bus?

   D3 D2 D1 D0

   __ __ __ __

2-5. In most computers, which approach permits the greater number of input/output circuits?
   a. Port-oriented input/output.
   b. Memory-mapped input/output.

2-6. In the circuit's address decoder, what type of logic element is used at C?
   a. AND.
   b. NAND.
   c. OR.
   d. NOR.
   e. XOR.

241

Chapter 18
INPUT/OUTPUT FOR SMALL COMPUTERS

Name _____

Date _____ Score _____

Instructor _____

INPUT/OUTPUT CIRCUIT DETAILS

3-1. The circuit shown below can be used as either an input or output. Ordinary flip-flops (not maste slave) are used, and their clocks are level triggered. When the clocks are active (low), number entered through their D leads immediately appear at the Q outputs. Also, the three-state bu fers can be effectively removed from the circuit by tying their enables to ground.

Use the drawing at the left to show the connections required to use this circuit as an output That is, connect PS (PORT SELECT) to A or B. Connect the unused enable (A or B) to ground

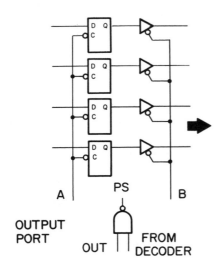

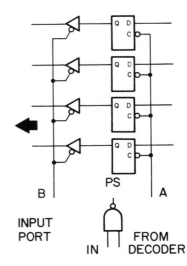

3-2. Repeat Problem 1 using the circuit at the right above. Show the connections necessary to us the circuit as an input.

3-3. The port-number decoder shown uses a 3-to-8 decoder chip. When a 3-bit binary number is applied to inputs CBA, the equivalent octal output lead goes active.

For each set of input signals, determine which port (0, 1, 2, or 3) will be activated. Write NONE if no port is activated by a given set.

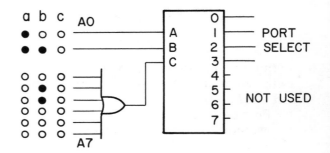

a. Port ____.

b. Port ____.

c. Port ____.

3-4. Match the following with the input/output types at the right.

A. Port oriented.
B. Memory mapped.
C. Direct memory access.

a. ____ Often referred to as accumulator input/output.

b. ____ Usually used to input/output large amounts of data.

c. ____ Input/output circuit treated as memory location.

d. ____ Microprocessor usually in standby mode during input/output.

Chapter 18
NPUT/OUTPUT FOR SMALL COMPUTERS

Name _____

Date _____ Score _____

Instructor _____

LIGHT-EMITTING DIODES

4-1. Circle the letters at the cathode leads on the displayed representations of LEDs. (Circle two letters.)

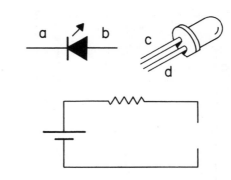

4-2. Add the symbol of an LED to the circuit at the right.

4-3. For the input signals shown, add switch arms to the switch equivalents at the right.

4-4. Which input from Problem 3 will cause the LED to light?
a. A.
b. B.
c. Both.

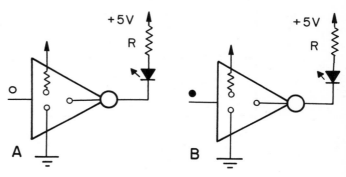

-5. Based on the circuit from Problem 3 and the equation at the right, determine the size of R that will result in a diode current of 15 mA. The 1.8 volts represents the voltage drop across the diode and allows 0.2 volts across the output of the logic element.

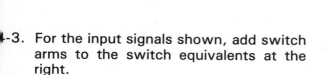

$$R = \frac{V_{cc} - 1.8}{I_d}$$

R = _____ Ohms.

-6. When the output of the displayed flip-flop is 1 (Q = 1), which LED will be lit?
a. A.
b. B.
c. Both.
d. Neither.

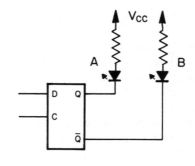

-7. For each circuit at the right, indicate the value of E (1 or 0) required to light the LED.

-8. Refer to the circuits from Problem 7. From the standpoint of heating the logic element, which is the preferred circuit?
a. A.
b. B.

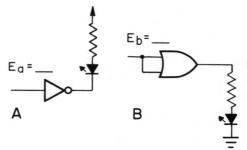

Chapter 18
INPUT/OUTPUT FOR SMALL COMPUTERS

Name _____

Date _____ Score _____

Instructor _____

LED CHARACTERISTICS

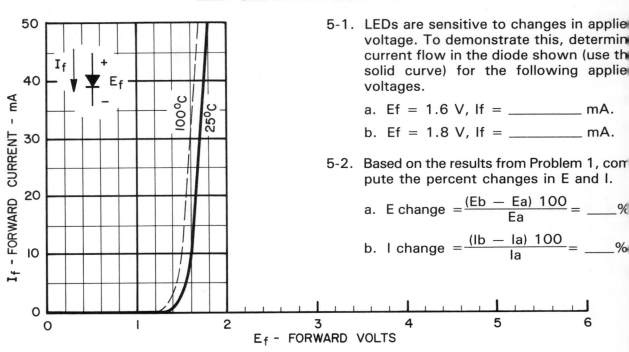

5-1. LEDs are sensitive to changes in applied voltage. To demonstrate this, determine current flow in the diode shown (use the solid curve) for the following applied voltages.

    a. E_f = 1.6 V, I_f = _____ mA.

    b. E_f = 1.8 V, I_f = _____ mA.

5-2. Based on the results from Problem 1, compute the percent changes in E and I.

    a. E change $= \dfrac{(E_b - E_a)\,100}{E_a} = $ ____%

    b. I change $= \dfrac{(I_b - I_a)\,100}{I_a} = $ ____%

5-3. LED current also changes with temperature. Characteristic curves for 100 °C and 25 °C (room temperature) are shown above. For an applied voltage of 1.6 volts, determine the currents for these temperatures.

    a. If at 25 °C = _____ mA.

    b. If at 100 °C = _____ mA.

The following discussion tells how to get good performance from LEDs. Current limiting resistors stabilize the operation of LED circuits. In the circuit at the right, for example, a 12.5% change in Vcc results in about a 20% change in If. While this change is large, it is smaller than the change you measured in Problem 1. Also, the resistor stabilizes the circuit from the standpoint of temperature. For a temperature change from 25 °C to 100 °C, If will increase about 1 mA.

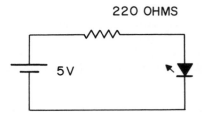

220 OHMS

5 V

5-4. Advanced Problem. If you have studied load lines in another course, construct load lines for the above circuit for Vcc = 5 V and Vcc = 5.6 V. Determine the following currents. Compare these results with those determined earlier on this sheet.

    a. If (5 V, 25 °C) = _____ mA.

    b. If (5 V, 100 °C) = _____ mA.

    c. If (5.6 V, 25 °C) = _____ mA.

Chapter 18

INPUT/OUTPUT FOR SMALL COMPUTERS

Name _____

Date _____ Score _____

Instructor _____

LEDs (LED, 220 OHM, 7404, VOLTMETER, AMMETER)

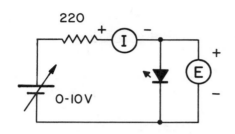

I_f – mA	E_f – VOLTS
1	
3	
5	
10	
20	
30	

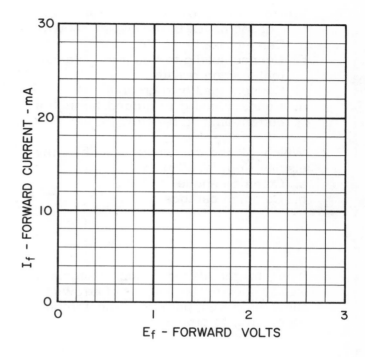

E_f – FORWARD VOLTS

5-1. Laboratory Activity. Use the above circuit to obtain data for the characteristic curve of the LED supplied by your instructor. Take data for the indicated currents. Plot your results on the graph. Estimate the average voltage across the diode for the range of currents used in the experiment.

Ef(av) = _____ V.

5-2. Laboratory Activity. Use the average voltage determined in Problem 1 and the equation at the right to predict the LED current in the circuit shown. Then build the circuit and test your prediction.

$$I_f = \frac{V_{cc} - E_f(av) - 0.2}{R}$$

a. If predicted = _____ mA.

b. If measured = _____ mA.

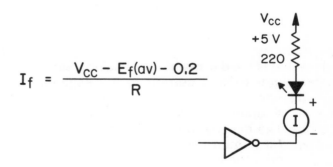

5-3. Laboratory Activity. Repeat Problem 2 for the circuit and equation shown. The 1 volt in this equation represents the drop across the transistor and diode in the output of the NOT. Due to differences in NOTs, your accuracy will be low.

$$I_f = \frac{V_{cc} - E_f(av) - 1}{R_o}$$

a. If predicted = _____ mA.
b. If measured = _____ mA.

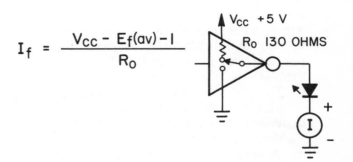

Chapter 18
INPUT/OUTPUT FOR SMALL COMPUTERS

Name _____

Date _____ Score _____

Instructor _____

SEVEN-SEGMENT DISPLAYS

7-1. Which of the following is the proper way to light segment g of the display shown?
   a. Connect lead g to Vcc.
   b. Connect lead g to Vcc through a current-limiting resistor.
   c. Connect lead g to ground.
   d. Connect lead g to ground through a current-limiting resistor.

7-2. Which of the following describes the display from Problem 1?
   a. Common anode.
   b. Common cathode.

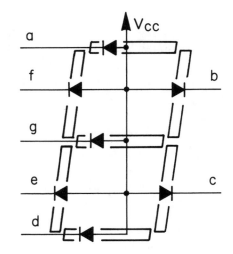

7-3. Refer to the decoder and 7-segment display shown. Use a series of arrows to show the current path from Vcc, through segment a, to ground.

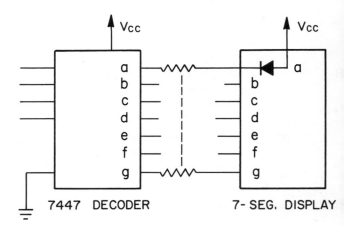

7-4. What is the disadvantage of using a single resistor to limit current to a 7-segment display?
   a. A single resistor will overheat.
   b. The decimal point will not light.
   c. All numbers will not have the same brightness.

7-5. Refer to the decoder with bubbled outputs and the display with bubbled inputs. Although seldom shown, it is permissible to use bubbles at the outputs of the decoder and inputs of the display from Problem 4. Why?
   a. To light a common-anode display, its inputs must be grounded.
   b. With bubbles at both input and output, their actions cancel.
   c. When used in this manner, the circles represent terminals.

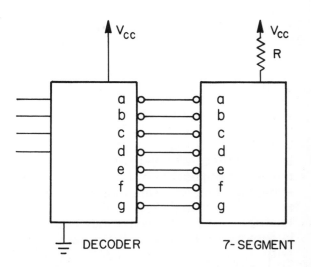

Chapter 18
INPUT/OUTPUT FOR SMALL COMPUTERS

Name _____

Date _____ Score _____

Instructor _____

SEVEN-SEGMENT DISPLAYS (MAN-1, 7-220 OHM, 7447)

18-1. Laboratory Activity. The drawing at the left shows power connections for a MAN-1. If a different display is used, your instructor will indicate the proper power connections. Note that some pins are missing. Although empty, these positions are counted when determining pin numbers.

Using a wire to ground individual pins, determine the internal connections of this display. DO NOT GROUND POWER LEADS. Use the resulting information to complete the pinout at the right. Mark missing pins NC (no connection) and power leads Vcc.

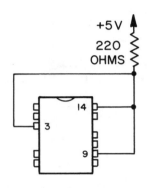

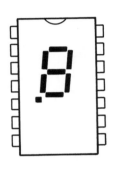

18-2. Laboratory Activity. Assemble the following and test its action. Input binary numbers 0000 through 1111. Between 0000 and 1001, record the decimal number displayed. Between 1010 and 1111, record the resulting patterns in the spaces provided.

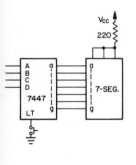

INPUT	0000	0001	0010	0011	0100
OUTPUT	____	____	____	____	____
	0101	0110	0111	1000	1001
	____	____	____	____	____

1010	1011	1100	1101	1110	1111
☐	☐	☐	☐	☐	☐

18-3. Laboratory Activity. Place a piece of paper over the display from Problem 2 such that only segments b and c are visible. Input 0001 through DCBA. Then ground test lead LT (this lights all segments). Note the change in brightness of b and c. Discuss the implications of using only one resistor.

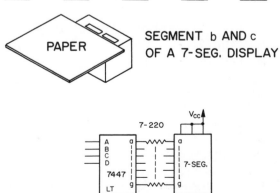

SEGMENT b AND c
OF A 7-SEG. DISPLAY

18-4. Laboratory Activity. Refer to the circuit with seven resistors. Rewire the circuit with a 220 Ohm resistor in series with each segment. Remove the resistor from the supply lead. Repeat Problem 3 for the circuit.

Chapter 18
INPUT/OUTPUT FOR SMALL COMPUTERS

Name _____

Date _____ Score _____

Instructor _____

MULTIPLEXED DISPLAY (7476, 3-7400, 2-MAN-1, 2-2N2222, 7447, 2-120 OHM, 2-1K)

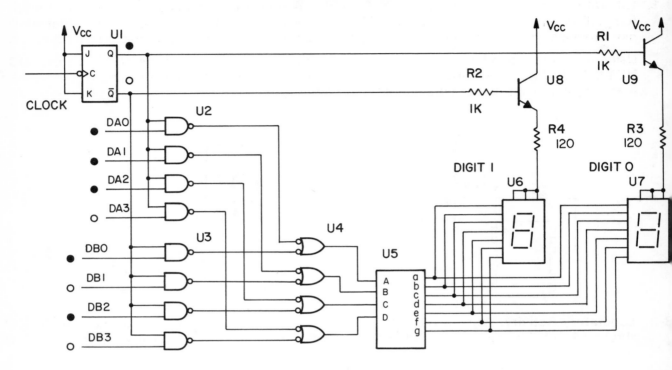

Problems on this sheet refer to the above multiplexed display circuit.

9-1. Match the following with the functions at the right.

    a. _____ U1.

    b. _____ U2, U3, U4.

    c. _____ U5.

    d. _____ U6, U7.

    e. _____ U8, U9.

A. Display.
B. Switch.
C. Decoder.
D. Frequency divider.
E. Multiplexer.

9-2. At the instant shown, which display is lit?
    a. Digit 1.
    b. Digit 0.
    c. Both are lit.
    d. Neither is lit.

9-3. Which number will be displayed at digit 0 when it is lit?
    a. 7 (from DA).
    b. 5 (from DB).

9-4. Laboratory Activity. If this is an experiment, build the circuit and test its action. Although can be built on one socket, it is better to use two. Permanently wire the number 0111 into inpu DA. Use switches for input DB. Operate the circuit at both high and low clock rates.

Chapter 18
INPUT/OUTPUT FOR SMALL COMPUTERS

Name _____

Date _____ Score _____

Instructor _____

KEYPAD WITH ENCODER (7411, 7486, 7432)

The following circuit represents a small keypad and its encoder. The truth table shows the circuit's action. When a single key is pressed, its number (0 through 3) is output in binary form. Also, lead KP (KEY PRESSED) goes active. The problems on this sheet refer to this circuit.

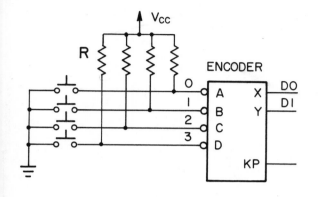

INPUTS				OUTPUTS		
A	B	C	D	Y	X	
0	1	2	3	DI	DO	KP
0	1	1	1	0	0	1
1	0	1	1	0	1	1
1	1	0	1	1	0	1
1	1	1	0	1	1	1

OTHER INPUT SETS RESULT IN ALL OUTPUTS BEING 0.

10-1. When key 2 (the switch in lead C) is pressed, what numbers (1s and 0s) will appear at the encoder's inputs?

a. Lead A = _____.

b. Lead B = _____.

c. Lead C = _____.

d. Lead D = _____.

10-2. Using sum-of-products, write expressions for the outputs of the above encoder.

a. _____ = Y.

b. _____ = X.

c. _____ = KP.

10-3. Simplify the above expressions. Note the presence of XORs. Also, the use of X or Y as an input to KP is possible. Show your work.

a. _____ = Y.

b. _____ = X.

c. _____ = KP.

10-4. Laboratory Activity. Implement the above expressions. If this is an experiment, build the circuit and test its action. Input signals can be obtained from the logic switches on your trainer.

10-5. When no key is pressed, outputs Y and X are 0 (D1D0 = 00). How does a computer attached to this circuit know the difference between the 00 when no key is pressed and the 00 that results from the pressing of key 0?
  a. It cannot tell the difference.
  b. It would look at input A to see if it is 0.
  c. It would not accept any number until KP is active.

A B C D
0 1 2 3

Chapter 18
INPUT/OUTPUT FOR SMALL COMPUTERS

Name _____

Date _____ Score _____

Instructor _____

MULTIPLEXED KEYPAD

11-1. The next few sheets will aid in the analysis of the multiplexed keypad shown.

When a key is pressed, its row number (D1D0) and its column number (D2) are output. Note that their order is reversed on the drawing (D0D1D2). These numbers are latched. That is, when a key is released, its number continues to be output.

Eight clock cycles are used to scan the pad. During the first three cycles, column 0 is scanned. The switch to column 1 is made on the fourth cycle. No keys are tested during this cycle. The next three cycles are used to inspect the switches in column 1. During cycle eight, the circuit returns to column 0. Again, no tests are made during this switching cycle.

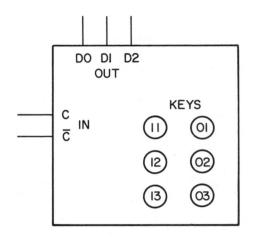

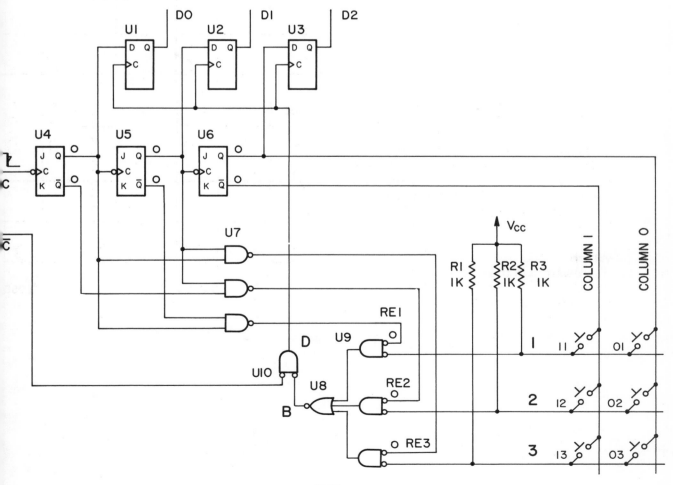

Chapter 18
INPUT/OUTPUT FOR SMALL COMPUTERS

Name _____

Date _____ Score _____

Instructor _____

MULTIPLEXED KEYPAD

The following problems refer to the circuit used with Problem 11-1 (called Problem 18-11-1).

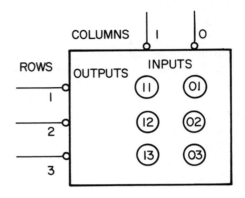

12-1. The drawing at the right shows the inputs and outputs of the keypad portion of the circuit used with Problem 11-1 (called Problem 18-11-1). The column select leads are active-low. The outputs from the rows are also active-low.

Only a key in a column with an active column enable can output an active signal. Show this by filling in the proper dots at the outputs of the circuits at the right. Only the key that is being pressed has been shown. Assume the other switches are open.

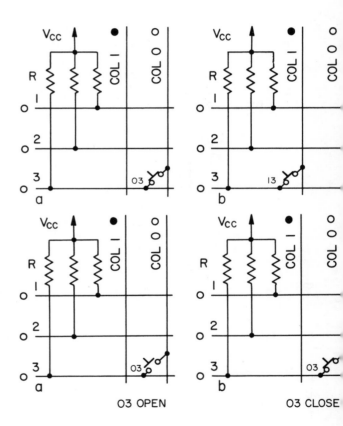

12-2. Assume that key 03 is being tested in the drawing at the right. For each state, indicate the outputs on the three row leads. (These outputs are active-low. To emphasize this, the NORs of U9 have been DeMorganed from NORs into elements with "AND" outputs. See the circuit used with Problem 11-1 [called Problem 18-11-1 or "Sheet 18-11"]).

12-3. Which of the following best describes the resistors in the circuit from Problem 2?
   a. Current-limiting.
   b. Pullup.
   c. Voltage divider.

12-4. Match the following flip-flop circuits on Sheet 18-11 with the functions at the right.

A. Ripple counter.
B. Ring counter.
C. Register.
D. Shift register.

    a. _____ U1, U2, U3.

    b. _____ U4, U5, U6.

12-5. Which flip-flop must be of the master-slave type?
    a. U1.
    b. U4.
    c. Both must be master-slave.
    d. Neither must be master-slave.

Chapter 18
INPUT/OUTPUT FOR SMALL COMPUTERS

Name _____

Date _____ Score _____

Instructor _____

MULTIPLEXED KEYPAD

13-1. Complete the displayed timing diagram to show the action of the ripple counter made up of U4, U5, and U6. Assume all Qs are 0 at t = 0.

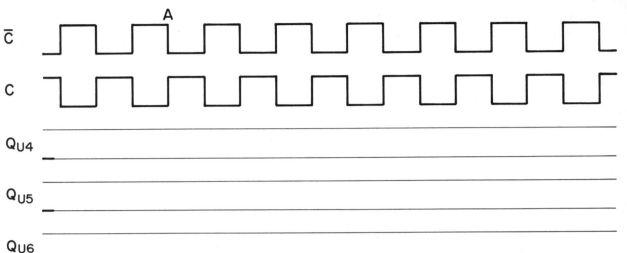

13-2. On the sheet used with Problem 11-1 (called Problem 18-11-1), dots appear at the output of the flip-flops that make up the ripple counter. By filling in one dot at the outputs of each flip-flop, indicate the count at time A in the above timing diagram. Also record that count the following spaces.

    a. QU4 = _____.

    b. QU5 = _____.

    c. QU6 = _____.

13-3. Based on the number in the counter at time A, which column of the keypad is active at this time
    a. COLUMN 1.
    b. COLUMN 0.

13-4. Based on the number in the counter at time A, determine the numbers (1s and 0s) at the three row-enable leads (RE1, RE2, and RE3). Fill in the proper dots at the inputs of the NORs U9 and also indicate these numbers in the following spaces.

    a. RE1 = _____.

    b. RE2 = _____.

    c. RE3 = _____.

13-5. Based on the answers to Problems 3 and 4, which key is being tested at time A?
    a. 11.   d. 02.
    b. 01.   e. 13.
    c. 12.   f. 03.

13-6. For each of the following conditions of the key selected in Problem 5, what number will appear at point B (the output of the NOR at U8)?

    a. Tested key being pressed — Point B = _____.

    b. Tested key not pressed — Point B = _____.

Chapter 18
INPUT/OUTPUT FOR SMALL COMPUTERS

Name _____

Date _____ Score _____

Instructor _____

MULTIPLEXED KEYPAD (2-7476, 2-7474, 7400, 7402, 7427 or ANOTHER 7402, 3-1K)

Problems on this sheet refer to the circuit used with Problem 11-1 (called Problem 18-11-1) and the timing diagram given with Problem 13-1 (called Problem 18-13-1).

18-1. On the sheet used with Problem 11-1 (called Problem 18-11-1), locate output Q of flip-flop U6. At time A, a 0 should appear at this point. Darken the path of this 0 through lead COLUMN 0, the keypad, the multiplexer, and the NOR at U10 to the clock inputs of the register (U1, U2, U3).

18-2. For the register to store the number at the output of the counter, what signal must appear at point D (the output of U10)?
a. Active-high.
b. Positive-going edge.
c. Active-low.
d. Negative-going edge.

18-3. The displayed timing diagram shows the signals applied to the NOR at U10 when key 02 is pressed. Complete the graph for lead D.

18-4. To indicate when the output of the counter is stored in the register, circle the proper letter under the graph for C.

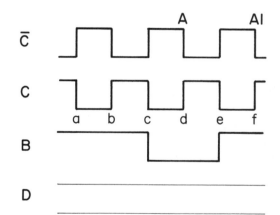

18-5. For the conditions in the timing diagram of Problem 3, what number will appear at the output of the register at time A1?
a. D0D1D2 = 010.
b. D0D1D2 = 111.
c. Whatever number appeared before key 02 was pressed.

18-6. Laboratory Activity. If this is an experiment, build and test the circuit used with Problem 11-1 (called Problem 18-11-1). Two sockets should be used. If two trainers are used, do not use both power supplies. This circuit should be built and tested in sections. The suggested order is: U4, U5, U6; U7; keypad; U9, U8, U10; and finally U1, U2, U3.

The circuit at the right can be used in place of a keypad. Connecting a column wire to one of the three row outputs will have the same effect as pressing a key.

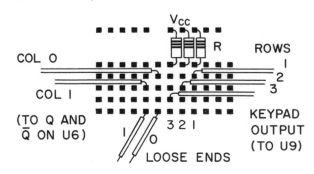

Chapter 19

COMPUTER ORGANIZATION AND MICROPROCESSORS

Name _____

Date _____ Score _____

Instructor _____

LANGUAGES

1-1. When the displayed program is run, what will appear on the screen? Assume the number 12 is entered for A and 23 for B. The instructions do the following tasks:

CLS — Clears screen.

PRINT — Places information on screen. Previously stored values for A, B, and C are displayed. Whatever appears between quote marks ('' '') is also displayed.

C = A+B — The results of this computation are stored as C.

END — Tells computer that the end of the program has been reached.

The semicolon (;) permits numbers entered through the keyboard to appear on the same line as the request for data. For example, instructions at 20 and 30 result in:

ENTER A . . . 12

```
10   CLS
20   PRINT ''ENTER A . . .'';
30   INPUT A
40   PRINT ''ENTER B . . .'';
50   INPUT B
60   C = A+B
70   PRINT A'' + ''B'' = ''C
80   END
```

1-2. Match the following with the programs at the right.

a. ____ High-level language.

b. ____ Most programs written in this form.

c. ____ Generally easier to write programs in this form.

d. ____ Can run directly on computer without being translated by INTERPRETER or COMPILER.

A.
```
10   CLS
20   PRINT ''PROGRAM 190''
30   PRINT ''ENTER PASSWORD'';
40   INPUT A$
50   etc.
```

B.
Memory Location		Number Stored	
04	82	0110	0100
04	83	0000	0110
04	84	0000	1000
		etc.	

1-3. Laboratory Activity. If this is an experiment, your instructor will supply instructions on programming a computer in your laboratory. Write, enter, and run programs to do the following:

    a. Compute the expected current when a voltage E is applied to a resistor R.
       The equation is: $I = E/R$.

    b. Compute the parallel resistance R of any two resistors A and B.
       The equation is: $R = A*B/(A + B)$.

Chapter 19
COMPUTER ORGANIZATION AND MICROPROCESSORS

Name _____

Date _____ Score _____

Instructor _____

MACHINE-LANGUAGE INSTRUCTIONS

Memory Location (Hex.)		Stored Number (Binary)
00	00	1001 0001
00	01	0000 1010
00	02	0000 0011
03	0A	0111 1100

The above numbers have been stored in RAM. The function of the instruction is to bring a number from memory and place it in the MPU's accumulator. The address of the number to be transferred is stored as shown below.

00	00	Op code, LOAD A from memory
00	01	Low-order byte of address
00	02	High-order byte of address

2-1. Four memory reads are required to execute the instruction. Use the drawings at the right to indicate the numbers on the three buses during these four reads. Control leads MEMORY READ and MEMORY WRITE are active-high.

2-2. When the execution of the instruction is complete, what number will be in the accumulator?

A = _____.

2-3. Where will the computer look for its next instruction?
a. 00 03.
b. 03 0A.
c. 03 0B.

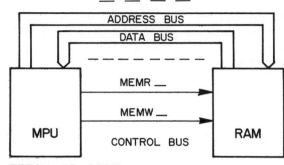

FETCH OP CODE

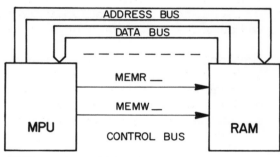

FETCH LOW-ORDER BYTE

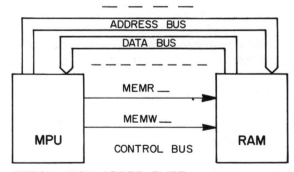

FETCH HIGH-ORDER BYTE

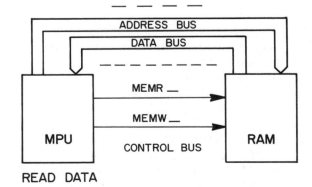

READ DATA

Chapter 19
COMPUTER ORGANIZATION AND
MICROPROCESSORS

Name_____

Date _____ Score _____

Instructor _____

MACHINE-LANGUAGE PROGRAMMING

Refer to the program on the page with the identifying label at the top in the form "PROGRAM STEP given as 'SHEET 19-5'." Assume that the program is stored in the RAM of the computer show at the top of that sheet. When RUN is activated, the computer goes to 00 00 and executes the instruction found there. Spaces in the program emphasize individual instructions. They do not appear in memory.

3-1. When RUN is activated, the instruction at the triplet 00 00, 00 01, and 00 02 is executed. Where does the computer look for its next op code?
   a. 00 03.
   b. 00 A0.
   c. 03 08.
   d. 03 B0.

3-2. When the instruction at 03 08, 03 09, and $A = _____.$
03 0A has been completed, what number
will be in the accumulator (in binary)?

3-3. How does the computer know that the number at 00 A0 is data?
   a. It does not. It merely does what the instruction tells it to do—place the number in memory location 00 A0 in the accumulator.
   b. A number like 0011 0100 is always data.
   c. Data is normally placed right after the first instruction, so the computer knows where look for data.

3-4. Normally, will the instruction at 03 08, 03 09, and 03 0A set flags?
   a. Yes (it involves the accumulator).
   b. No (it is not an arithmetic/logic instruction).

3-5. What number will be in the accumulator $A = _____.$
when the instruction at 03 0B, 03 0C,
and 03 0D has been completed (in
binary)?

3-6. Based on the answer to Problem 5, indicate the states (1s and 0s) of the following flags.

   a. ZERO = _____.

   b. SIGN = _____.

   c. CARRY = _____.

   d. PARITY = _____.

3-7. Based on the instruction at 03 0E, 03 0F, and 03 10, what will cause the conditional jump to take place?
   a. The number in register C is 0000 0001.
   b. The CARRY FLAG is set.
   c. The sum (called C) from the addition equals 0000 0001.

3-8. Which of the following is the probable purpose for testing C with a conditional jump?
   a. To see if the sum extends into register C.
   b. To be sure the accumulator did not clear.
   c. To see if an overflow occurred.

Chapter 19
COMPUTER ORGANIZATION AND
MICROPROCESSORS

Name _____

Date _____ Score _____

Instructor _____

MACHINE-LANGUAGE PROGRAMMING
(Refer to the Program Steps on Sheet 19-5.)

19-1. Based on the data in memory, where will the computer look for its next op code after completing the instruction at 03 0E, 03 0F, and 03 10?
a. 00 00.
b. 03 08.
c. 03 11.
d. 03 B0.

19-2. What number will be in the accumulator when the execution of the instruction at 03 11, 03 12, and 03 13 is complete?

A = _ _ _ _ _ _ _ _ _.

19-3. Based on the results from Problem 2, indicate the state (1s and 0s) of the following flags.

a. ZERO = _____.

b. SIGN = _____.

c. CARRY = _____.

d. PARITY = _____.

19-4. When the execution of the instruction at 03 14, 03 15, and 03 16 is complete, where will the computer look for its next op code?

a. 00 00.

b. 03 08.

c. 03 17.

d. 03 B0.

19-5. What number will be in the accumulator when the execution of the instruction at 03 B0 and 03 B1 is complete?

A = _ _ _ _ _ _ _ _ _.

19-6. Refer to the output instruction stored at 03 B2 and 03 B3. Use the following drawing to indicate the numbers on the three buses at the time the output is taking place. Let the higher-order byte of the address bus be 00. All control leads are active-high.

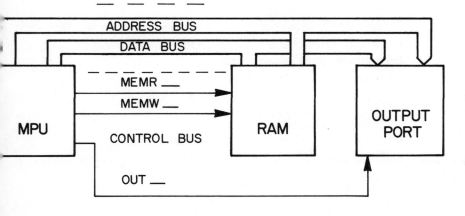

Note that the program has been *ABORTED* (ended early). This was caused by the overflow. The outputting of FF (hexadecimal) through port 04 is an error message. It tells the operator what has happened. If there had been no overflow, the program would have continued.

Chapter 19
COMPUTER ORGANIZATION AND
MICROPROCESSORS

Name _____

Date _____ Score _____

Instructor _____

PROGRAM STEPS Given as "SHEET 19-5"

The displayed program is referred to on the sheets for Problem 3-1 through Problem 3-8, Problem 4-1 through Problem 4-6, and Problem 6-1 through Problem 6-6.

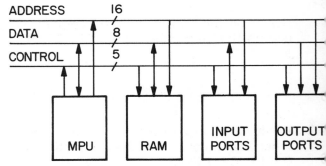

Memory Location (Hex.)		Stored Number (Binary)		
00	00	0000	1100	Op code, UNCONDITIONAL JUMP
00	01	0000	1000	Low-order byte of address of jump
00	02	0000	0011	High-order byte of address of jump
00	A0	0011	0100	Data
00	A1	1001	0100	Data
00	A2	0100	0010	Data
03	08	1001	0001	Op code, LOAD A from memory
03	09	1010	0000	Low-order byte of address
03	0A	0000	0000	High-order byte of address
03	0B	0111	0100	Op code, ADD memory to A
03	0C	1010	0001	Low-order byte of address
03	0D	0000	0000	High-order byte of address
03	0E	0110	0000	Op code, CONDITIONAL JUMP if C = 1
03	0F	1011	0000	Low-order byte of address
03	10	0000	0011	High-order byte of address
03	11	0111	0100	Op code, ADD memory to A
03	12	1010	0010	Low-order byte of address
03	13	0000	0000	High-order byte of address
03	14	0110	0000	Op code, CONDITIONAL JUMP if C = 1.
03	15	1011	0000	Low-order byte of address
03	16	0000	0011	High-order byte of address
03	17	0101	1100	Op code, OUTPUT number in A
03	18	0000	0011	Port number for output
03	19	0000	1100	Op code, UNCONDITIONAL JUMP
03	1A	1011	0100	Low-order byte of address
03	1B	0000	0011	High-order byte of address
03	B0	1100	0011	Op code, MOVE next byte to A
03	B1	1111	1111	Number to be moved to A
03	B2	0101	1100	Op code, OUTPUT number in A
03	B3	0000	0100	Port number for output
03	B4	1111	1111	Op code, HALT

Name _____

Date _____ Score _____

Instructor _____

MACHINE-LANGUAGE PROGRAMMING

19-1. Assume that the program on Sheet 19-5 contained the data at the right. Would it run (no overflow) or abort (overflow)?
   a. Run.
   b. Abort.

00	A0	1100	0010
00	A1	0011	1111
00	A2	0001	0000

19-2. Repeat Problem 1 for the second data list.
   a. Run.
   b. Abort.

00	A0	1001	0110
00	A1	0101	0001
00	A2	0001	0010

19-3. If the error-detecting instruction (CONDITIONAL JUMP if C = 1) had not been used, what would happen when an overflow occurred?
   a. The program would run and output faulty sums.
   b. The program would stop when the overflow occurred, but no error message would be sent.

19-4. How many memory-read cycles are needed to execute each of the following instructions?

   a. _____ LOAD A from memory (03 08, 03 09, and 03 0A).

   b. _____ CONDITIONAL JUMP if C = 1 (03 0E, 03 0F, and 03 10).

   c. _____ MOVE next byte to A (03 B0 and 03 B1).

   d. _____ HALT (03 B4).

19-5. What would happen if the numbers in locations 00 01 and 00 02 were interchanged? (*Interchanged.)
   a. The program would run just as well with these numbers reversed.
   b. The first jump would be to 08 03, so the program would not run.
   c. The computer will detect the error and output an error message.

00	00	0000	1100
00	01	*0000	0011
00	02	*0000	1000

19-6. The numbers located at 03 B1 and 03 B4 are both 1111 1111. How does the computer tell the difference between them?

| 03 | B1 | 1111 | 1111 |
| 03 | B4 | 1111 | 1111 |

   a. It does not. They are both treated as op codes.
   b. The computer knows that the last instruction will be HALT, so it treats the last number in the program as an op code.
   c. The op code for MOVE next byte to A tells the computer that the number in the next memory location is just that, a number. In turn, when the computer completes the instruction at 03 B2 and 03 B3, it knows the number in the next memory location is an op code.

Chapter 19
COMPUTER ORGANIZATION AND
MICROPROCESSORS

Name _____

Date _____ Score _____

Instructor _____

MICROPROCESSOR BLOCK DIAGRAM

Problems on this sheet refer to the block diagram of an MPU at the right.

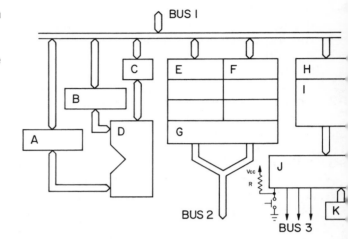

7-1. Match the external buses (external to the MPU) with the following functions.
   A. Address.
   B. Data.
   C. Control.

   a. _____ Bus 1.

   b. _____ Bus 2.

   c. _____ Bus 3.

7-2. Match the following functions with the blocks in the diagram.

   a. _____ Instruction register.

   b. _____ Instruction decoder.

   c. _____ Control and timing.

   d. _____ Program counter.

   e. _____ and _____ Accumulator and temporary register 3.

   f. _____ and _____ Temporary registers 1 and 2.

   g. _____ ALU.

   h. _____ Flag register.

   i. _____ Clock.

7-3. Place the letters B and C in the blocks that might represent REGISTERS B and C.

7-4. Control leads within the MPU have not been shown. Add the control lead that would be use
to signal the ACCUMULATOR that it is to latch a number that is on the DATA BUS.

7-5. The switch near block J is the RUN input. What is the function of the resistor?
   a. Pullup resistor.
   b. Current-limiting resistor.

7-6. In the computer depicted, is RUN active-high or active-low?
   a. Active-high.
   b. Active-low.

7-7. When RUN is activated, CONTROL AND TIMING places 0000,0000,0000,0000 (the addres
of the first op code) in which of the following registers or register pairs?
   a. Blocks A and B.
   b. Blocks E and F.
   c. Block G.

Chapter 19
COMPUTER ORGANIZATION AND
MICROPROCESSORS

Name _____

Date _____ Score _____

Instructor _____

MPU ACTION

The MPU of a small computer is shown below. Assume the following instruction is stored in RAM.

03	A4	1100 0011 (C3)	Op code, MOVE next byte to A
03	A5	0100 1100 (4C)	Number to be moved to A

19-1. When the fetch of the op code has been completed, what numbers will appear in the PC and INSTRUCTION REGISTER? Place your answers in the proper blocks in hexadecimal.

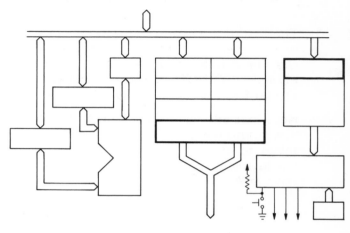

19-2. When the execution of the above instruction has been completed, what numbers will appear in ACCUMULATOR, PC, and INSTRUCTION REGISTER? Indicate your answers in hexadecimal on the diagram.

19-3. Will the execution of the instruction near the top of the page set flags?
a. No, it does not involve the ALU.
b. Yes, it changes the number in the accumulator.

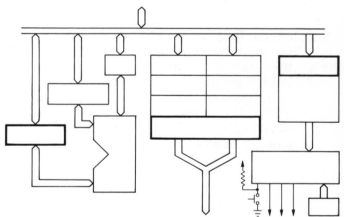

19-4. How did the MPU know that it had to fetch a second byte from memory to complete the instruction?
a. When the op code was decoded, it indicated that the next byte in memory must be brought in.
b. The second byte is always brought in.
c. The number in the second byte (0100 1100) indicated that it was data and had to be brought in.

19-5. Where will this computer look for its next op code?
a. 03 A5.
b. 03 A6.
c. 03 4C.
d. 00 00.

19-6. T   F   Data that might have been in the ACCUMULATOR before the execution of this instruction is lost (cannot be recovered). Circle the correct answer.

**Chapter 19
COMPUTER ORGANIZATION AND
MICROPROCESSORS**

Name _____

Date _____ Score _____

Instructor _____

MPU ACTION (Continuation of the Sheet for Problem 8-1 through Problem 8-6

03	A4	1100 0011 (C3)	Op code, MOVE next byte to A
03	A5	0100 1100 (4C)	Number to be moved to A
03	A6	0111 0110 (76)	Op code, ADD next byte to A
03	A7	0111 0000 (70)	Number to be added to A

9-1. The first of the above instructions places the number 4C in the ACCUMULATOR. Use Fig. A to indicate the numbers in the PC and INSTRUCTION REGISTER after the fetch of the op code at 03 A6.

Note that once a number is in a register, it will not change unless formal action is taken to change it. In this case, bringing in the op code at 03 A6 does not change A.

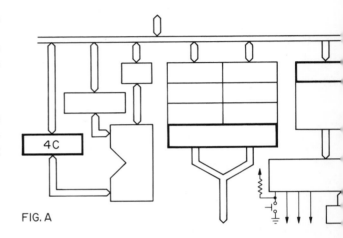

FIG. A

9-2. Repeat Problem 1, called Problem 19-9-1, for the time just after the completion of the fetch of the number to be added to A but before the actual addition. Use Fig. B to indicate the numbers in ACCUMULATOR, TEMPORARY 3, INSTRUCTION REGISTER, and PC. Assume the PC has already been incremented.

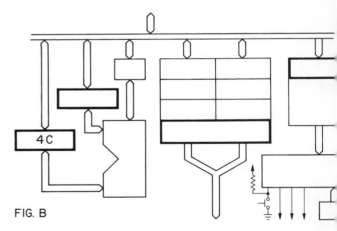

FIG. B

9-3. Repeat Problem 1, called Problem 19-9-1, for the time after the addition has been completed. Use Fig. C to indicate the numbers in ACCUMULATOR, TEMPORARY REGISTER 3, INSTRUCTION REGISTER, and PC. (PC would not be incremented by this action.)

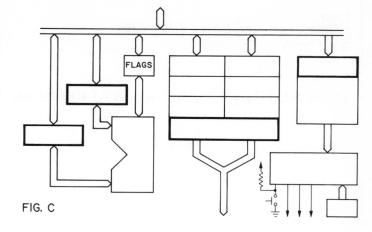

FIG. C

9-4. Indicate the states (1s and 0s) of the following flags after the completion of the above instruction.

a. ZERO = _____.

b. SIGN = _____.

c. CARRY = _____.

d. PARITY = _____.

Chapter 19
COMPUTER ORGANIZATION AND
MICROPROCESSORS

Name _____

Date _____ Score _____

Instructor _____

MPU ACTION (Continuation of the sheet for Problem 9-1 through Problem 9-4

03	A4	1100 0011 (C3)	Op code, MOVE next byte to A
03	A5	0100 1100 (4C)	Number to be moved to A
03	A6	0111 0110 (76)	Op code, ADD next byte to A
03	A7	0111 0000 (70)	Number to be added to A
03	A8	0110 0000 (60)	Op code, CONDITIONAL JUMP if C = 1
03	A9	0000 1101 (0D)	Low-order byte of address
03	AA	0000 0100 (04)	High-order byte of address

10-1. Three fetches are needed to bring in the instruction at 03 A8, 03 A9, and 03 AA. Indicate the numbers in ACCUMULATOR, TEMPORARY REGISTERS 3, 1, and 2, INSTRUCTION REGISTER, and PC after the three fetches but before the final execution of the instruction. Use hexadecimal.

10-2. For the flag values found in Problem 4, called Problem 19-9-4, what number will be in the PC when the execution of the instruction is complete?

PC = __ __ __ __.

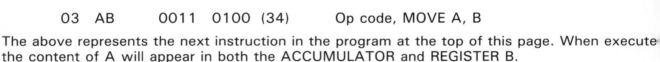

| 03 | AB | 0011 0100 (34) | Op code, MOVE A, B |

The above represents the next instruction in the program at the top of this page. When execute the content of A will appear in both the ACCUMULATOR and REGISTER B.

10-3. Indicate the numbers in ACCUMULATOR, TEMPORARY REGISTER 3, REGISTER B, INSTRUCTION REGISTER, and PC when the execution of the instruction at 03 AB is complete.

Remember, the content of a register does not change unless formal action is taken to change it. That is, the number in A is not changed by the MOVE action.

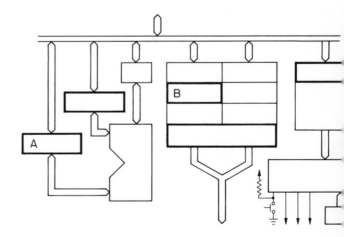

Chapter 20

TTL LOGIC ELEMENT SPECIFICATIONS AND OPERATING REQUIREMENTS

Name _____

Date _____ Score _____

Instructor _____

TTL TOTEM-POLE OUTPUT CIRCUIT

1-1. In the group of three transistors shown, which transistor is likely to be conducting?

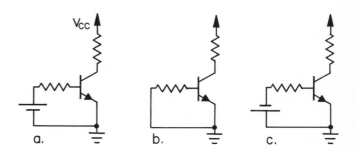

a.   b.   c.

1-2. For the transistor you selected in Problem 1, draw and label two arrows. One is to show the path of the base current (Ib); the other is to show the path of the collector current (Ic) from Vcc to ground.

1-3. Which switch equivalent represents the state of the totem-pole circuit at the right?

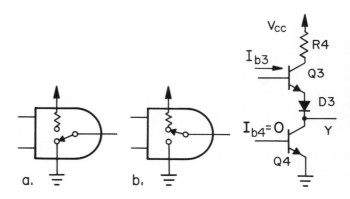

a.   b.

1-4. Which switch equivalent most closely represents the action of a totem-pole output circuit?

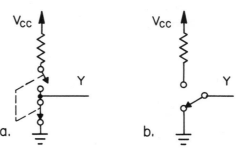

a.   b.

269

1-5. Match the following functions with the components of the circuit at the right.

    a. ____ Switch that grounds output when 0 is to be output.

    b. ____ Switch that connects Y to Vcc when 1 is to be output.

    c. ____ Part of bias circuit.

    d. ____ Limits output current.

Chapter 20
TTL LOGIC ELEMENT SPECIFICATIONS
AND OPERATING REQUIREMENTS

Name _____

Date _____ Score _____

Instructor _____

TTL CIRCUIT OF THE TYPE CALLED "STANDARD" (3-2N2222, 1N914, 1K, 1.8K, 10K, 120 OHMS)

2-1. In the circuit shown, I1 is flowing, so Q2 is conducting. Use a series of arrows to show the path (or paths) of the collector current Ic2 of Q2 from Vcc to ground.

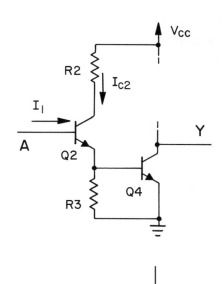

2-2. How large must the base-to-emitter voltage of the displayed transistor be to cause it to conduct?
a. 0.3 V.
b. 0.7 V.
c. 1.4 V.
d. 5 V.

2-3. In the circuit shown, I1 is flowing, so Q2 and Q4 are on. Their base-to-emitter voltages are about 0.7 V. Their collector-to-emitter voltages are about 0.2 V. As a result, the indicated voltages (with respect to ground) can be expected.

Assume that voltage difference between points B and Y divides equally between the base-to-emitter junction of Q3 and the diode D3. What is the base-to-emitter voltage of Q3?

Ebe3 = _____ V.

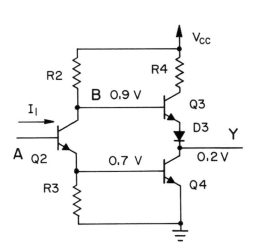

2-4. Based on the answer to Problem 3, will Q3 be conducting?
a. Yes.
b. No.

2-5. Laboratory Activity. If this is an experiment, built this NOT and test its action. With A = 1 (I1 flowing), measure and record the indicated voltages with respect to ground. Based on the base-to-emitter voltages of the output transistors, circle the one that is conducting. Due to differences in transistors, measured voltages will vary.

Retain this circuit. It will be used in the next experiment.

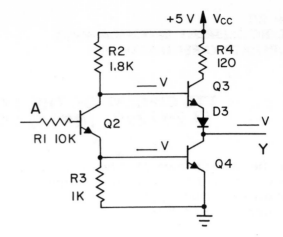

Chapter 20
**TTL LOGIC ELEMENT SPECIFICATIONS
AND OPERATING REQUIREMENTS**

Name _____

Date _____ Score _____

Instructor _____

TTL CIRCUIT OF THE TYPE CALLED "STANDARD"
(3-2N2222, 1N914, 1K, 1.8K, 10K, 120 OHMS)

3-1. Here, $I1 = 0$, so transistors Q2 and Q4 are off. To emphasize this, these transistors have been omitted in the second drawing.

RL represents the load placed on the circuit by the input of the element being driven.

Without the flow of I2 (the collector current of Q2), there is little voltage drop across R2. As a result, the voltage between point B and ground is about 4.8 V. The output voltage is about 3.4 V.

Assume that the voltage difference between points B and Y divides equally between the base-to-emitter junction of Q3 and the diode D3. What is the base-to-emitter voltage of Q3?

Ebe3 = _____ V.

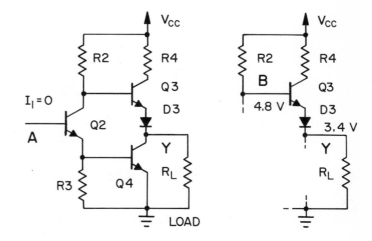

3-2.   T   F   Based on the voltage in Problem 1, Q3 will be conducting. Circle the correct answer.

3-3.   Two currents flow in Q3—its base current and its collector current. Use two sets of arrows to trace the paths of these currents from Vcc to ground. Use the circuit at the left above.

3-4.   When does diode D3 serve its function in the above circuit?
a. When Q3 is conducting.
b. When Q3 is turned off.

3-5.   Laboratory Activity. If this is an experiment, build the circuit shown and observe its output for A = 1 and A = 0. Then, with A = 0 (I1 is not flowing), measure the indicated voltages with respect to ground. Based on the base-to-emitter voltages, which output transistor (Q3 or Q4) is conducting when A = 0?

Retain this circuit. It will be used in a later experiment.

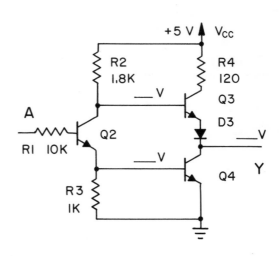

Chapter 20
TTL LOGIC ELEMENT SPECIFICATIONS
AND OPERATING REQUIREMENTS

Name _____

Date _____ Score _____

Instructor _____

TTL DECISION-MAKING (4-2N2222, 1N914, 1K, 1.8K, 4.7K, 120 OHMS

4-1. Add arrows to the following circuits to show the path (or paths) of I1 through Q1.

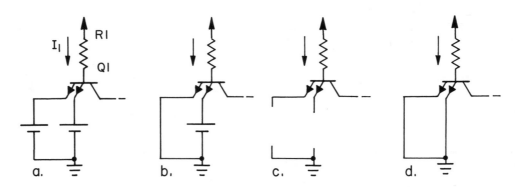

4-2. For the logic 0 being input at the right, trace the path of I1 to ground. This circuit is a NOT.

4-3. For the current flow, indicated in Problem 2, circle all transistors that are conducting. If I1 flows from the base to the emitter of Q1, it is conducting. If I1 flows from the base to the collector of Q1, it is not conducting.

4-4. Based on the transistors circled in Problem 3, what signal appears at output Y?
a. 1.
b. 0.

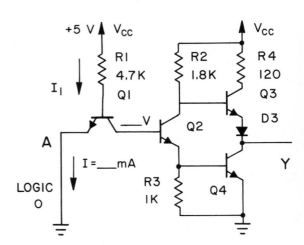

4-5. For the logic 1 being input at the right, trace the path of I1 to ground.

4-6. Repeat Problem 3 for A = 1. Use the circuit from Problem 5.

4-7. Repeat Problem 4 for A = 1.
a. Y = 1.
b. Y = 0.

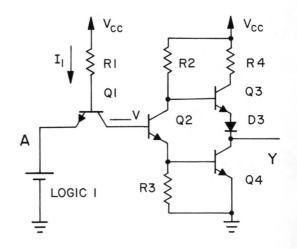

4-8. Laboratory Activity. If this is an experiment, build the circuit at the top of this sheet, connec it to Q2, and test its action. Measure the voltage between the base of Q2 and ground. Whe Q2 is off, this voltage should be near 0. When Q2 is on, it should be near 1.4 V. Also measur the current flowing out of the input when A = 0. Record your results on the drawings.

Chapter 20
TTL LOGIC ELEMENT SPECIFICATIONS
AND OPERATING REQUIREMENTS

Name _____

Date _____ Score _____

Instructor _____

TTL INPUT/OUTPUT OF THE "STANDARD" TYPE

5-1. In the circuit shown, a TTL output is driving two TTL inputs. The driving element is outputting a 0. Use arrows to show the paths of the currents I11 and I12 to ground.

5-2. Assume I11 and I12 are both at their maximum values (1.6 mA). How much current will Q4 in Problem 1 have to sink?

I4 = _____ mA.

5-3. In standard TTL elements, what is the current rating of Q4?

I4 max. = _____ mA.

5-4. What is the fanout of standard TTL outputs?

Fanout = _____.

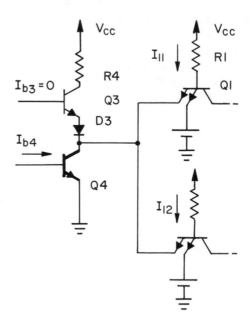

5-5. When the emitter on Q1 at the right is turned off, a small reverse current flows (less than 40 microamperes). Use a series of arrows to show the path of Ir from Vcc, through the output circuit of the driving element, and Q1.

5-6. Based on a fanout of 10, what must be the current rating of Q3?

I3 recommended max. = _____ microamperes.

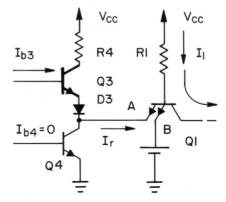

5-7. Which output signal will result in the flow of current in the short circuit shown at the right?
a. Logic 0.
b. Logic 1.

5-8. Which of the following best represents the current a standard TTL output will deliver to a short circuit?
a. 4 mA.
b. 40 mA.
c. 400 mA.

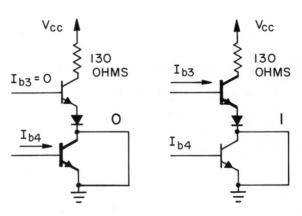

Chapter 20
TTL LOGIC ELEMENT SPECIFICATIONS
AND OPERATING REQUIREMENTS

Name _____

Date _____ Score _____

Instructor _____

TTL CIRCUITS OF THE "STANDARD" TYPE

6-1. Which of the following best describes the circuit shown?
   a. Buffer/driver.
   b. Three-state NAND.
   c. Open-collector NAND.

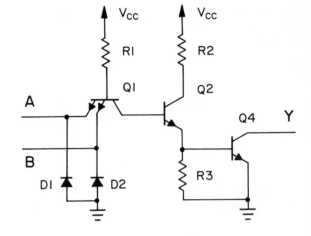

6-2. Which of the following switch equivalents best represents the above circuit?

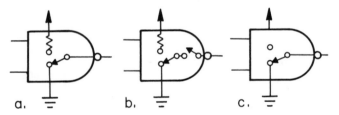

6-3. Match the following functions with the letters and numbers on the schematic shown. If there is more than one correct answer, give only one.

   a. ____ Input lead.

   b. ____ Source of power to drive circuit.

   c. ____ Switch that grounds output when Y = 0.

   d. ____ Switch in decision-making circuit.

   e. ____ Diode that supplies bias.

   f. ____ Clamping diode.

   g. ____ Resistor that limits output current.

   h. ____ Diode that speeds action of element.

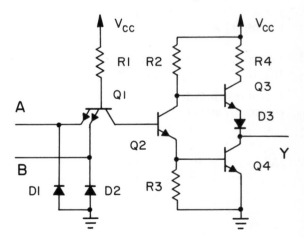

Chapter 20
TTL LOGIC ELEMENT SPECIFICATIONS
AND OPERATING REQUIREMENTS

Name _____

Date _____ Score _____

Instructor _____

TTL SPECIFICATIONS FOR THE "STANDARD" TYPE

20-1. The displayed graph shows the specified output and input voltages of a standard TTL NOT. Complete this graph by placing voltages in the eight blanks.

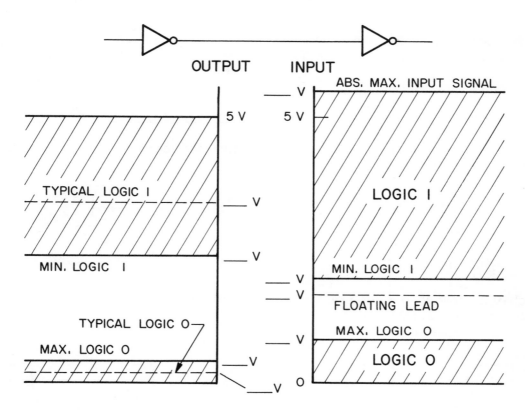

OUTPUT   INPUT

20-2. The drawing of a mechanical device represents the adjustment on a power supply used on a standard TTL circuit. Into which range should the voltage supplied to such a circuit fall?
a. 3.00-7.00 V.
b. 4.25-5.75 V.
c. 4.75-5.25 V.
d. 4.99-5.01 V.

INCREASE

V_{CC}

20-3. On the average, how much current and power is drawn by a standard TTL NAND element? The term "current/gate" means "current per gate."

a. Average current/gate = _____ mA.

b. Average power/gate = _____ mW.

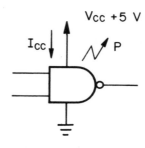

Chapter 20
TTL LOGIC ELEMENT SPECIFICATIONS
AND OPERATING REQUIREMENTS

Name _____

Date _____ Score _____

Instructor _____

TTL SPECIFICATION AND CIRCUITS OF THE "STANDARD" TYPE

8-1. Determine the propagation delay depicted in the graph at the right.

tP = _____ ns.

8-2. Which of the following best represents the propagation delay of standard TTL NANDs?
   a. 10 ns.
   b. 60 ns.
   c. 120 ns.

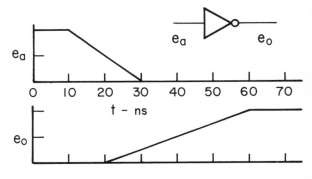

8-3. Which of the following is not a satisfactory method for dealing with unused leads on ORs and NORs in a laboratory?

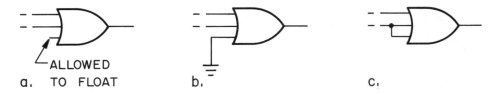

a. TO FLOAT   b.   c.

8-4. Which of the following is not a satisfactory method for dealing with unused leads on ANDs and NANDs in a laboratory?

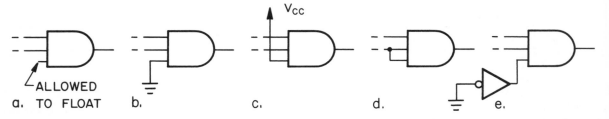

a. TO FLOAT   b.   c.   d.   e.

8-5. In the following space, draw the schematic for a 2-input NAND from memory. Label all components (Q1, D1, R1, etc.). Label the inputs A and B and the output Y. Indicate Vcc.

Chapter 21

LOGIC FAMILIES OTHER THAN STANDARD TTL FAMILY

Name _____

Date _____ Score _____

Instructor _____

RELATIONSHIP OF POWER AND SPEED

-1. Which logic element is dissipating the most power?
a. A.
b. B.

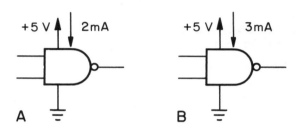

-2. On which board are the benefits of low power likely to be important?
a. A.
b. B.

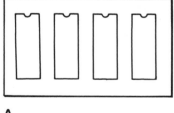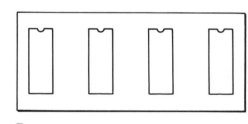

A   B

-3. Determine the propagation delay represented by the displayed graph.

tP = _____ ns.

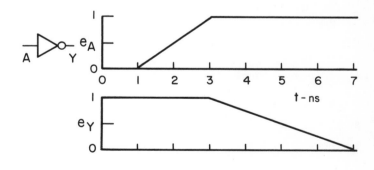

-4. Which family will have flip-flops with the higher clock frequency rating?
a. One with short propagation delay.
b. One with long propagation delay.

1-5. The displayed NOTs are from different families. Their relative propagation delays can be determined from the graphs. Which is likely to dissipate the most power?
   a. Chip 1.
   b. Chip 2.

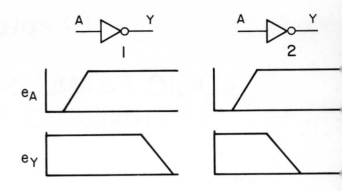

Chapter 21
LOGIC FAMILIES OTHER THAN
STANDARD TTL FAMILY

Name _____

Date _____ Score _____

Instructor _____

LOW POWER TTL SERIES OF LOGIC ELEMENTS

21-1. Ohm's Law states that the current (in amperes) through a resistor equals the voltage across that resistor divided by its resistance (in Ohms). Use Ohm's Law to compute the current through the resistors shown.

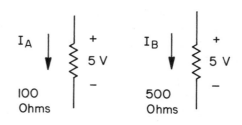

a. IA = _____ A.
b. IB = _____ A.

$$I = \frac{E}{R} \qquad P = EI$$

21-2. Watt's law states that the power (in watts) delivered to a resistor equals the product of the voltage across it times the current (in amperes) through it. Based on the results of Problem 1, compute the power dissipated by each resistor.
a. PA = _____ Watts.
b. PB = _____ Watts.

21-3. Based on the results of Problem 2, complete the following statement by selecting the correct word.

For a given applied voltage, the resistor with the (a) larger (b) smaller resistance will dissipate the SMALLER power.

21-4. Which circuit is most likely low power TTL?

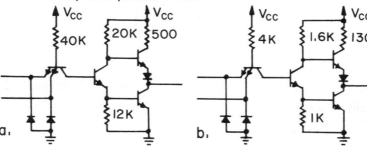

21-5. Currents, at the instant the switches were closed, are shown at the right. Which circuit will probably take the longer time to charge its capacitor?
a. A.
b. B.

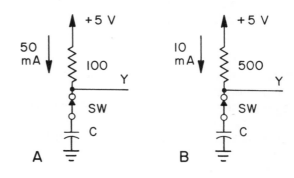

21-6. Based on the answer to Problem 5, which circuit probably has the longer propagation delay?
a. A.
b. B.

Chapter 21
LOGIC FAMILIES OTHER THAN
STANDARD TTL FAMILY

Name _____

Date _____ Score _____

Instructor _____

LOW POWER TTL SERIES OF LOGIC ELEMENTS

3-1. Place the following letters in the blanks to indicate the relation between standard TTL and the 74LXX series of logic elements.

L = Considerably lower.
H = Considerably higher.
E = About equal.

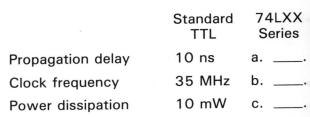

	Standard TTL	74LXX Series
Propagation delay	10 ns	a. _____.
Clock frequency	35 MHz	b. _____.
Power dissipation	10 mW	c. _____.

3-2. Low power signal currents are shown at the right. Based on these numbers, what is the fanout of the series?

Fanout = _____.

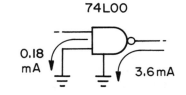

3-3. The elements in the circuit shown are all standard TTL. Would it be permissible to replace the NAND with a 74L00?
a. Yes.
b. No.

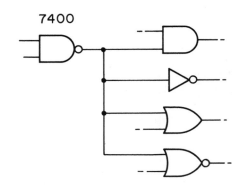

3-4. What is the probable function of the 74L04 in the circuit shown?
a. To decrease the fanin of the 7404.
b. To invert the output of the MPU.
c. To add necessary delay into the memory read control lead.

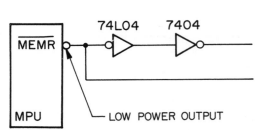

3-5. Based on the information available in the displayed circuit, would it be acceptable to replace the 74L04 with a 7404? (The other five NOTs in the chip would have to be checked to see if the change would affect their operation. In this case, however, base your answer only on this circuit.)
a. Yes.
b. No.

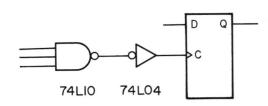

Name _____

Date _____ Score _____

Instructor _____

SATURATED/NONSATURATED OPERATION OF TRANSISTORS

4-1. The displayed curve shows the relationship between the base and collector currents of the transistor with the current
arrows and collector resistor. At which
base current is the transistor saturated?
a. A.
b. B.

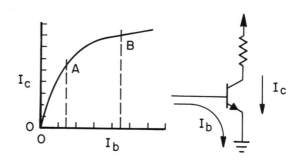

4-2. The switch equivalents shown represent
saturated and nonsaturated transistors.
Which drawing most likely represents
saturated operation?
a. A.
b. B.

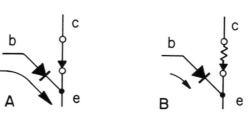

4-3. Based on current flowing through the switch portion of the symbols in Problem 2, which type
of operation results in LOWER power dissipation?
a. Saturated.
b. Nonsaturated.

4-4. What type of operation is used in the low power TTL series of logic elements? (This type of
operation is also used in the standard TTL series.)
a. Saturated.
b. Nonsaturated.

4-5. Although saturated operation results in
low power dissipation, it has disadvan-
tages. In saturated transistors, charge
carriers are stored in their bases. These
must be removed before saturated transistors can turn off. The resulting effect
called *STORAGE TIME* adds to the propagation delay. Which element is most likely
operating with saturated transistors?
a. Chip 1.
b. Chip 2.

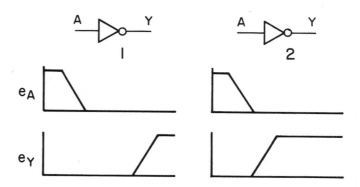

4-6. To avoid long storage time, high-speed families of logic elements usually operate nonsaturated.
How does this effect the power dissipation of such families?
a. It lowers power requirements.
b. It increases power requirements.

Chapter 21
LOGIC FAMILIES OTHER THAN
STANDARD TTL FAMILY

Name _____

Date _____ Score _____

Instructor _____

SCHOTTKY TTL SERIES OF LOGIC ELEMENTS

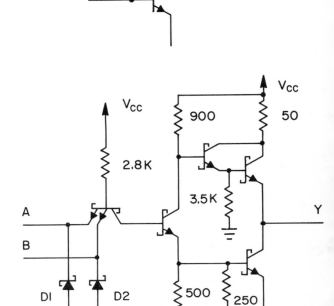

5-1. How does the Schottky diode keep the displayed transistor out of saturation?
a. It bypasses excess current around the base.
b. It rectifies the incoming signal.
c. It applies a negative voltage to the base.

5-2. Which of the following is not a method of producing high-speed operation in the circuit shown?
a. Schottky transistors have been used.
b. Resistances have low values.
c. There is no diode in the totem pole portion of the circuit.
d. Extra drive stages have been added.

5-3. What effect has increasing the speed of the circuit from Problem 2 had on the power dissipated?
a. Increased it.
b. Decreased it.

5-4. Place the following letters in the blanks to indicate the relation between standard TTL and the 74SXX series.

L = Considerably lower.
H = Considerably higher.
E = About equal.

	Standard TTL	74SXX Series
Propagation delay	10 ns	a. ____
Clock frequency	35 MHz	b. ____
Power dissipation	10 mW	c. ____

5-5. Flip-flops from the 74SXX series are used in the frequency divider shown. It divides an 80 MHz signal to 20 MHz. Could standard 74XX series flip-flops be substituted in this circuit (consider frequency and what drives the first input)?
a. Yes.
b. No.

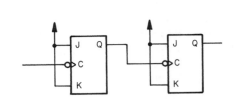

Chapter 21
LOGIC FAMILIES OTHER THAN
STANDARD TTL FAMILY

Name _____

Date _____ Score _____

Instructor _____

DIODE LOGIC (4-1N914, 2-1K, VOLTMETER)

6-1. Laboratory Activity. For the logic levels shown below, complete the PREDICTED column of the table. Then build the circuit and test your predictions. Record your results in the MEASURED column. Is the circuit a NAND, AND, OR, NOR, or NOT?

Logic 1 = Approximately +5 V.   Logic 0 = Approximately 0 V.

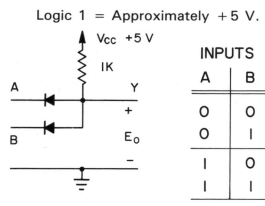

INPUTS		PREDICTED	MEASURED
A	B	Y	Y
0	0		
0	1		
1	0		
1	1		

6-2. Laboratory Activity. With Y = 1, measure the circuit's output voltage. Do not disassemble your circuit. It will be used later.

EO open = _____ V.

6-3. Laboratory Activity. Repeat Problem 1 for the circuit having two diodes with a resistor to ground. Is the circuit a NAND, AND, OR, NOR, or NOT?

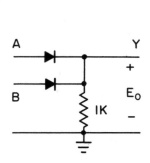

INPUTS		PREDICTED	MEASURED
A	B	Y	Y
0	0		
0	1		
1	0		
1	1		

6-4. Laboratory Activity. Repeat Problem 2 for the circuit in Problem 3.

6-5. Laboratory Activity. Diode logic does not provide amplification. As a result, logic levels decrease as signals pass from one element to another. To demonstrate this, connect the above AND and OR as shown. Measure Eloaded and Eout. Compare these voltages with the open-circuit voltages noted above.
Eloaded = _____ V.
Eout   = _____ V.

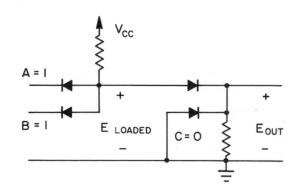

**Chapter 21
LOGIC FAMILIES OTHER THAN
STANDARD TTL FAMILY**

Name _____

Date _____ Score _____

Instructor _____

LOW POWER SCHOTTKY TTL SERIES OF LOGIC ELEMENTS

7-1. What form does the decision-making portion of the displayed circuit take?
   a. Totem pole.
   b. Multiple-emitter.
   c. Diode.
   d. ECL.

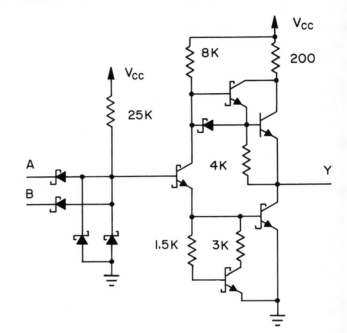

7-2. The following list describes methods used to produce high-speed circuits. Which is not used in the 74LSXX element from Problem 1?
   a. Schottky transistors.
   b. Low resistance values.
   c. Extra driving stages.

7-3. Place the following letters in the blanks to indicate the relation between standard TTL and the 74LSXX series of logic elements.

   L = Considerably lower.
   H = Considerably higher.
   E = About equal.

	Standard TTL	74LSXX SERIES
Propagation delay	10 ns	a. ____.
Clock frequency	35 MHz	b. ____.
Power dissipation	10 mW	c. ____.

7-4. The counter shown uses standard TTL flip-flops and operates at 32 MHz. On the basis of frequency, could 74LSXX chips be substituted?
   a. Yes.
   b. No.

7-5. 74LSXX signal currents are shown at the right. What is the fanout of this family?

   Fanout = _____ .

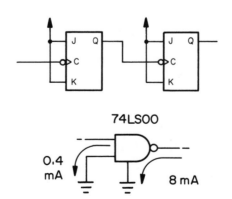

74LS00

7-6. The elements in the displayed circuit are all standard TTL. Based on signal current ratings, could a 74LS00 be substituted for the NAND? (The other three NANDs in the chip would have to be checked to see that such a substitution is possible. However, base your answer on this circuit only.)
a. Yes.
b. No.

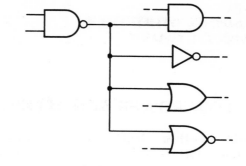

**Chapter 21
LOGIC FAMILIES OTHER THAN
STANDARD TTL FAMILY**

Name _____

Date _____ Score _____

Instructor _____

CMOS FAMILY OF LOGIC ELEMENTS

8-1. Refer to the CMOS output sections shown. For each input signal, circle the transistor or transistors that will be turned on. Also, indicate the resulting output signal (1 or 0).

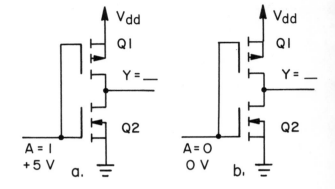

8-2. Repeat Problem 1 for the displayed circuits and input signals.

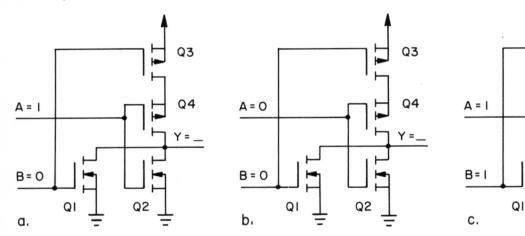

8-3. What logic element is represented by the circuit in Problem 2?
   a. AND.
   b. NAND.
   c. OR.
   d. NOR.

8-4. In the circuit shown, a NOT is driving a NOT. Starting at Vdd, use a series of arrows to show the flow of current through the circuit to ground. If there is no path, write NONE.

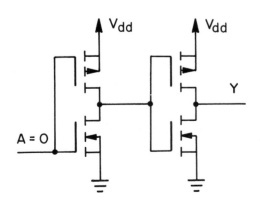

8-5. Based on your answer to Problem 4, which statement best describes the static power required by CMOS elements?
  a. Essentially zero.
  b. Medium amounts of power will be dissipated.
  c. Because Vdd is shorted to ground, dissipation will be high.

8-6. Place the following letters in the blanks to indicate the relation between standard TTL and the CMOS family.

  L = Considerably lower.
  H = Considerably higher.
  E = About equal.

	Standard TTL	CMOS Family
Propagation delay	10 ns	a. _____.
Clock frequency	35 MHz	b. _____.
Power dissipation	10 mW	c. _____.

Chapter 21
LOGIC FAMILIES OTHER THAN
STANDARD TTL FAMILY

Name _____

Date _____ Score _____

Instructor _____

CMOS AND ECL FAMILIES OF LOGIC ELEMENTS

9-1. What is the function of the diodes at the input of the CMOS circuit shown?
a. Reduces ringing.
b. Protects against static charges.
c. Provides bias for transistors.

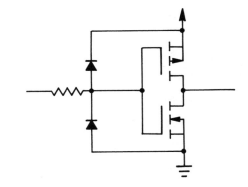

9-2. When do CMOS elements dissipate the most power?
a. When signals are changing (from 1 to 0 or 0 to 1).
b. When a logic level is applied for a long time.

9-3. ECL stands for emitter coupled logic. Circle the emitter resistor that accomplishes this coupling in the NOT at the right.

9-4. What is unique about the power supply voltage used in the ECL circuit from Problem 3?
a. Its voltage is much higher than required by other families.
b. It is negative with respect to ground.
c. It is an ac (alternating current) voltage.

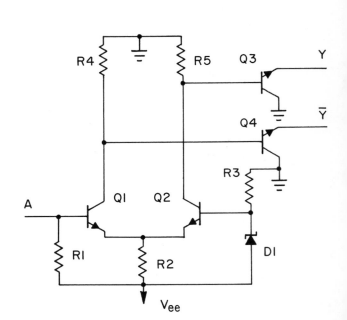

9-5. What form of output circuit is used in the ECL NOT from Problem 3?
a. Totem pole.
b. Open-collector.
c. Three-state.

9-6. Place the following letters in the blanks to indicate the relation between standard TTL and the ECL family.

L = Considerably lower.
H = Considerably higher.
E = About equal.

	Standard TTL	ECL Family
Propagation delay	10 ns	a. ____.
Clock frequency	35 MHz	b. ____.
Power dissipation	10 mW	c. ____.

Chapter 21
**LOGIC FAMILIES OTHER THAN
STANDARD TTL FAMILY**

Name _____

Date _____ Score _____

Instructor _____

FUNDAMENTALS OF LOGIC FAMILIES

10-1. Match the following with the families at the right. If there is more than one correct answer, give only one.

a. ____ Fastest.

b. ____ Most power per gate.

c. ____ Lowest power per gate.

d. ____ Slowest.

e. ____ A low power replacement for standard TTL.

f. ____ Most easily damaged by static electricity.

g. ____ Has totem-pole output.

h. ____ Signal voltage is negative with respect to ground.

i. ____ Highest fanout.

j. ____ Uses nonsaturated transistors.

k. ____ Most difficult to interface with standard TTL elements.

l. ____ Dissipates very little power when static signals are applied.

A. Standard TTL.
B. Low power TTL.
C. Schottky TTL.
D. Low power Schottky TTL.
E. CMOS.
F. ECL.

m. ____
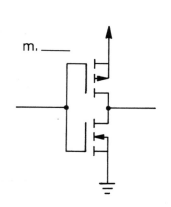

n. ____
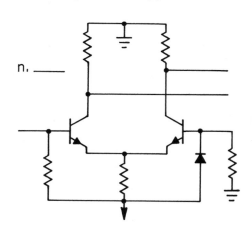

10-2. Which of the displayed circuits is a diode AND?

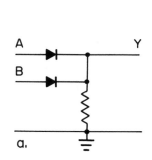

a.

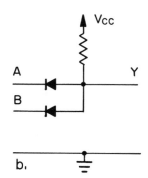

b.

Chapter 22

RELAY LOGIC AND PROGRAMMABLE CONTROLLERS

Name _____

Date _____ Score _____

Instructor _____

HOLDING CONTACTS

-1. Refer to the ladder diagram having two horizontal control lines. Which relay contacts act as holding contacts?
   a. 1CR.
   b. 2CR.
   c. 4CR.

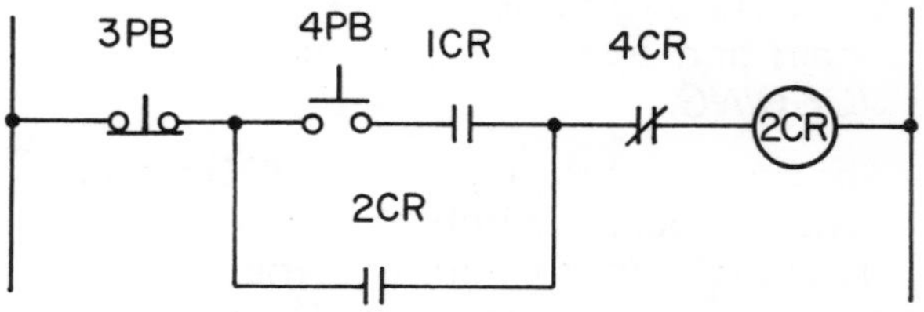

-2. Refer to the circuit from Problem 1. Indicate the condition of each of the devices necessary to energize relay 2CR. Do this by using the letters:

   A = Active.
   N = Not active.
   D = Don't care.

   a. _____ 3PB.

   b. _____ 4PB.

   c. _____ 1CR.

   d. _____ 4CR.

-3. Repeat Problem 2 for the time after 2CR has been energized. That is, indicate the conditions of the devices necessary to keep 2CR energized.

   a. _____ 3PB.

   b. _____ 4PB.

   c. _____ 1CR.

   d. _____ 4CR.

1-4. To light the lamp in the circuit shown, 2PB must be pressed first. The lamp will then light when 3PB is pressed. When STOP is pressed, both relays de-energize and the lamp will go out.

Which is the proper connection for contacts 2CR on line 4?
   a. A.
   b. B.
   c. Either will work.

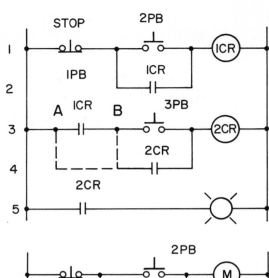

1-5. During setup and repair, it is sometimes necessary to move a machine forward very slowly. On some machines, this can be done by starting and stopping them again and again. This process is called *JOGGING.*

The control at the right has a JOG feature. When selector switch 1SS is in JOG, the machine's motor starts when 2PB is pressed and it stops when 2PB is released. When in RUN, the motor continues to run when 2PB is released.

Which is the JOG position of 1SS?
   a. A.
   b. B.

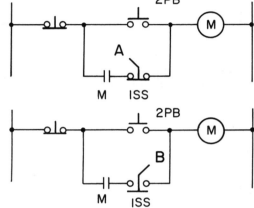

Chapter 22
RELAY LOGIC AND PROGRAMMABLE
CONTROLLERS

Name _____

Date _____ Score _____

Instructor _____

CONTROL CIRCUIT (Continued in Problem 3-1 through Problem 3-9 and Problem 4-1 through Problem 4-8.)

The displayed circuit controls the filling and emptying of a tank. When START/RESET is pressed, the tank fills. When full, the READY light comes on. An operator then presses DELIVER, and the pre-measured content of the tank is delivered to the next step in the process. During delivery, the DELIVERING light is on.

When empty, the tank is automatically refilled.

When STOP is pressed, the circuit is completely de-energized. START/RESET must be pressed to restart the process.

2-1. The following questions refer to the ladder diagram having 11 lines. Which push-button switch is normally open (NO)?
  a. 1PB.
  b. 2PB.

2-2. What is the meaning of the dashed line between the two parts of the symbol 2PB?
  a. The operator must press both switches at the same time.
  b. There is a mechanical link between the two sets of contacts, so they operate at the same time.

2-3. How many contacts on 1CR are used in the circuit?

  Number of contacts used = _____.

2-4. Which contacts on 1CR are holding contacts?
  Those on line(s) _____.

2-5. If STOP is pressed while 2CR (line 4) is energized, what will happen to this relay?
  a. It will de-energize due to the opening of 1CR on line 4.
  b. It will remain energized due to the holding action of 2CR on line 5.

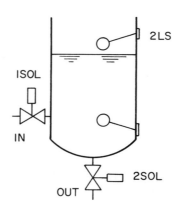

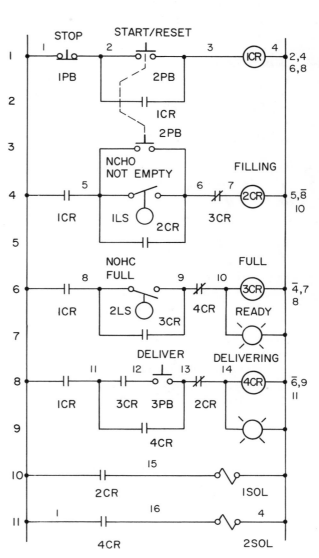

Chapter 22
RELAY LOGIC AND PROGRAMMABLE
CONTROLLERS

Name _____

Date _____ Score _____

Instructor _____

CONTROL CIRCUIT (Refer to the sheet for Problem 2-1 through Problem 2-5 Continued on the sheet for Problem 4-1 through Problem 4-8.)

3-1. On which line is the solenoid that operates the fill (inflow) valve located?

On line _____.

3-2. To energize the solenoid referred to in Problem 1 (called Problem 3-1), which relays must b
energized? Mark as many as needed.
a. 1CR.
b. 2CR.
c. 3CR.
d. 4CR.

3-3. What are the two ways of starting the filling process? Mark two.
a. Press START/RESET (2PB).
b. Close tank-empty limit switch (1LS).
c. Close tank-full limit switch (2LS).
d. Release DELIVER (3PB).

3-4. When the tank is empty, 1LS (line 4) closes. As soon as the tank begins to fill, however, 1L
again opens. How is 2CR kept energized?
a. The operator continues to press 2PB.
b. 3CR (line 4) closes.
c. 2CR (line 5) closes.
d. Relay 2CR (line 4) is a latching relay.

3-5. Which of the following senses that the tank is full?
a. 1LS (line 4).
b. 2LS (line 6).
c. 3PB (line 8).

3-6. How is the fact that the tank is full recorded electrically by the circuit?
a. 1CR is de-energized.
b. 3CR (line 6) is energized.
c. 4CR (line 8) is energized.
d. 2SOL (line 11) is energized.

3-7. When the tank is full, solenoid 1SOL must be de-energized. That is, 2CR must be de-energize
How is this accomplished?
a. The operator watches the tank and presses STOP.
b. 1LS opens and turns 2CR off.
c. Contacts 3CR on line 4 open and turn 2CR off.

3-8. When the tank is full and ready to deliver, which relays will be energized? Mark as many as necessary.
   a. 1CR.
   b. 2CR.
   c. 3CR.
   d. 4CR.

3-9. Before delivery, but still under the conditions of Problem 8 (called Problem 3-8), which solenoid or solenoids will be energized?
   a. 1SOL.
   b. 2SOL.
   c. Neither.
   d. Both.

Chapter 22
RELAY LOGIC AND PROGRAMMABLE
CONTROLLERS

Name _____

Date _____ Score _____

Instructor _____

CONTROL CIRCUIT (Refer to the sheet for Problem 2-1 through Problem 2-5.

4-1. If DELIVER (3PB on line 8) were pressed before the tank was full (before 3CR had been ene
gized), what would happen?
   a. Nothing, since 3CR on line 8 would be open.
   b. The filling would stop because this button caused the outputting of the liquid.
   c. The filling would continue, but liquid would be flowing in and out at the same time.

4-2. When will the light marked READY (line 7) be on?
   a. Whenever 3CR is energized.
   b. Only while 2LS is closed.
   c. When 3PB on line 8 is released.

4-3. When the tank is full (3CR is energized) and DELIVER (3PB) is pressed, which of the followin
will NOT happen?
   a. 3CR will de-energize.
   b. 4CR will energize.
   c. The ready lamp (line 7) will go out.
   d. The delivering lamp (line 9) will go out.
   e. Solenoid 2SOL (line 11) will energize.

4-4. When the tank is empty, 1LS (line 4) closes and energizes 2CR. How is the output soleno
2SOL (line 11) turned off?
   a. 2CR on line 8 opens and turns off 4CR on line 8.
   b. 2CR turns on 1SOL (line 10), and it turns off 2SOL.
   c. 2SOL is not turned off at this point in the cycle.

4-5. If STOP is pressed during the delivery (that is, while 4CR is energized), what would happer
   a. Delivery would stop, because 1CR on line 8 would open and de-energize 4CR.
   b. Delivery would continue, because the holding contacts on line 9 would keep 4CR energize
   c. Delivery would continue even though 4CR would be de-energized, because there are no 1C
   contacts on line 11.

4-6. On which lines are the contacts of relay 3CR located?

   On lines _____.

4-7. What numbers will be on the wires leading to the following contacts?

   a. Limit switch 1LS on line 4: _____ and _____.

   b. 1CR contacts on line 8:   _____ and _____.

4-8. T   F   The circuit has undervoltage protection. Circle the correct answer.

Chapter 22
RELAY LOGIC AND PROGRAMMABLE CONTROLLERS

Name _____

Date _____ Score _____

Instructor _____

MOTOR-STARTING CIRCUIT

Problems on this sheet refer to the circuit at the right.

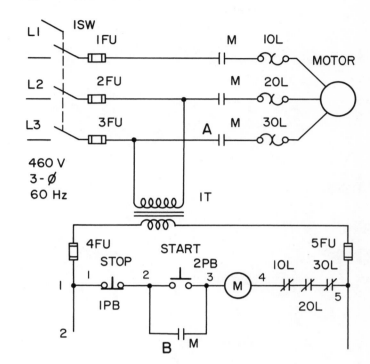

5-1. Draw a solid line around the power portion of the circuit shown. Draw a dashed line around the control portion.

5-2. What name is given to transformer 1T?
   a. Power transformer.
   b. Control transformer.
   c. Isolation transformer.

5-3. What voltage probably appears across the ladder portion of the circuit?
   a. 5 V.
   b. 120 V.
   c. 460 V.

5-4. How many contacts on contactor M are used in this circuit?

   Number of contacts on M = _____.

5-5. Which contacts on M are likely to be designed to carry the most current?
   a. Those at A.
   b. Those at B.

5-6. What is the function of switch 1SW?
   a. To start and stop the motor.
   b. To remove power during circuit repair, etc.

5-7. If a massive short circuit were to occur across the motor leads, which protective devices would probably protect the service?
   a. 1FU 2FU 3FU.
   b. 4FU 5FU.
   c. 1OL 2OL 3OL.

5-8. If a small, long-term overload caused the motor to overheat, which protective devices would probably activate?
   a. 1FU 2FU 3FU.
   b. 4FU 5FU.
   c. 1OL 2OL 3OL.

5-9. Why would it be wrong to save the cost of the transformer by connecting the control circuit to a 120 V source?

a. Undervoltage protection would be lost.

b. Such voltages are ac; control circuits are dc.

c. Sufficient power would not be available.

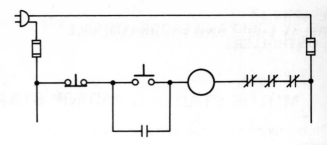

Chapter 22
RELAY LOGIC AND PROGRAMMABLE CONTROLLERS

Name _____

Date _____ Score _____

Instructor _____

MOTOR-STARTING CIRCUIT (Continued on the Sheet for Problem 7-1 through Problem 7-7.)

During starting, electric motors draw higher than normal currents. To limit these currents, resistors 1R, 2R, and 3R have been placed in the motor leads of the circuit shown. When the motor is up to speed (its current is back to normal), these resistors are shorted out of the circuit by contacts R. Time-delay relay TR determines how long these resistors are in the circuit.

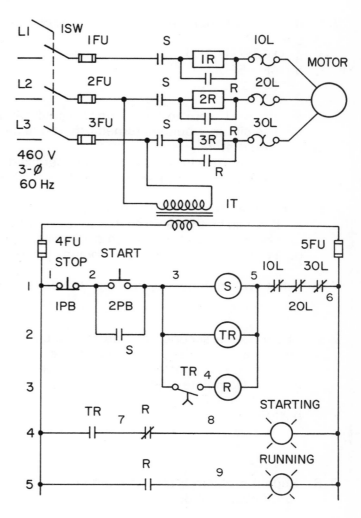

7-1. Based on the above description, which set of contacts in the power portion of the circuit closes first during the starting process?
   a. S.
   b. R.

7-2. The arrow on the contacts labeled TR (line 3) indicates that the delay is on the closing of these contacts. That is, when the coil of TR is energized, these contacts do not close until the time delay has run out. When START is pressed, which coil is NOT energized immediately?
   a. S.
   b. TR.
   c. R.

7-3. When is lamp STARTING lit?
   a. When S and R are energized.
   b. When TR and R are energized.
   c. When TR is energized, but R is de-energized.

7-4. How many contacts on contactor R are used in this circuit?

Number of contacts on R used = _____.

7-6. T   F   The circuit has undervoltage protection. Circle the correct answer.

7-6. If power is lost, what happens to the contactors and time delay relay?
   a. They all de-energize, and the starting process must be repeated to restart the motor.
   b. S and TR are de-energized, but R remains energized. As a result, the restarting process is shorter after a power outage.
   c. All three coils remain energized, so the motor automatically starts when power returns.

Chapter 22
RELAY LOGIC AND PROGRAMMABLE
CONTROLLERS

Name _____

Date _____ Score _____

Instructor _____

MOTOR-STARTING CIRCUIT (Refer to the circuit on the sheet for Problem 6-1 through Problem 6-6.)

7-1. If the coil of contactor S were to short circuit, which group of protective devices would probabl operate?
  a. 1FU 2FU 3FU.
  b. 4FU 5FU.
  c. 1OL 2OL 3OL.

7-2. If fuse 5FU were to blow, what would happen?
  a. The circuit would be turned off. The relay and contactors would de-energize, and the mot would stop.
  b. The control circuit would be de-energized, but the motor would continue to run until stoppe by the opening of switch 1SW.

7-3. If overload relay 1OL were to open, what process would be used to restart the motor?
  a. It would be the same as that for a cold start. START would be pressed, and the current limitir resistors would remain in series with the motor until the time delay ran out.
  b. When the overload relay cooled, the motor will start itself.

7-4. What wire numbers appear on the leads to contacts TR on line 3?

   Wire numbers for TR = _____.

7-5. Show how a second start button would be connected in the circuit shown. Pressing either start button should start the motor. Complete the circuit between wires 1 and 3. Label the second start button 3PB.

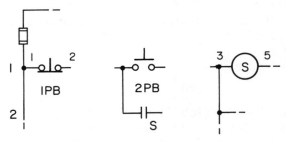

7-6. Repeat problem 5 for a second stop button. Label it 4PB.

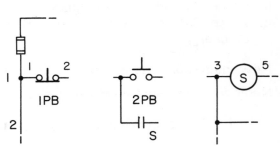

7-7. If START and STOP are pressed at the same time, what happens?
  a. The motor starts.
  b. The motor will not start.

**Chapter 22
RELAY LOGIC AND PROGRAMMABLE
CONTROLLERS**

Name _____

Date _____ Score _____

Instructor _____

INDUSTRIAL CONTROL COMPONENTS

8-1. The displayed pushbutton switch has two
sets of contacts. Which is NC?

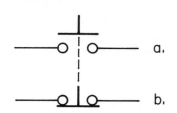

8-2. On ladder diagrams, switches are shown in the positions they are in at the beginning of the
machine's cycle. Which of the following shows a normally-open limit switch that is being held
closed?

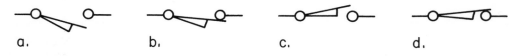

a.   b.   c.   d.

8-3. Match the following with the symbols shown below. If there is more than one correct answer,
give only one. A given symbol may be used more than once.

a. ____ Fuse.

b. ____ Overload relay.

c. ____ Limit switch.

d. ____ Indicator lamp.

e. ____ Control relay.

f. ____ Pushbutton switch.

g. ____ Control transformer.

h. ____ Three-phase induction motor.

i. ____ Contactor.

j. ____ Disconnect switch.

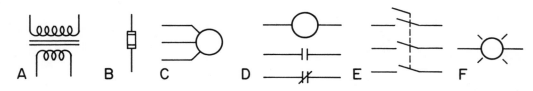

8-4. A typical motor-starting circuit can be drawn. Use the above symbols to complete the motor starting circuit drawing provided. It is to have undervoltage protection and overload protection. Label all components.

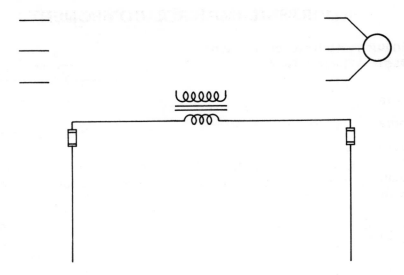

Chapter 22
RELAY LOGIC AND PROGRAMMABLE
CONTROLLERS

Name _____

Date _____ Score _____

Instructor _____

PROGRAMMABLE CONTROLLERS

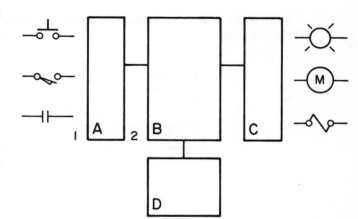

9-1. Match the following with the sections of the programmable controller system shown.

    a. ____ Programmable controller.

    b. ____ Programming console.

    c. ____ Input modules.

9-2. At which point in the block diagram of Problem 1 is the signal voltage likely to be +5 V?
    a. Point 1.
    b. Point 2.

9-3. Which of the following best describes a programmable controller?
    a. A group (array) of logic elements that can be interconnected (hardwired) to act like the relays and switches used in industrial controls.
    b. A computer that can be programmed to simulate the actions of relay logic.

9-4. Which of the following is the starting point for the programming of a programmable controller?
    a. Ladder diagram.
    b. Logic diagram.
    c. Boolean expression.

For True and False questions, circle the correct answer.

9-5.   T   F   It usually costs less to program a programmable controller than to construct and wire a relay panel.

9-6.   T   F   It normally costs less to make changes in a programmable controller program than to make wiring changes in a relay panel.

9-7.   T   F   Programmable controllers interface easily with computers.

9-8.   T   F   Modern programmable controllers are capable of both analog and digital control.

9-9.   T   F   Because they contain computers, programmable controllers can do arithmetic operations.

9-10. T   F   When a machine is scrapped, its relay panel can usually be salvaged and used on another machine. This is not true of programmable controllers.

Chapter 22
RELAY LOGIC AND PROGRAMMABLE
CONTROLLERS

Name _____

Date _____ Score _____

Instructor _____

ANALOG TYPE OF PROGRAMMABLE CONTROLLERS

10-1.　Which of the displayed drawings represents or involves an analog signal or an analog relationship?

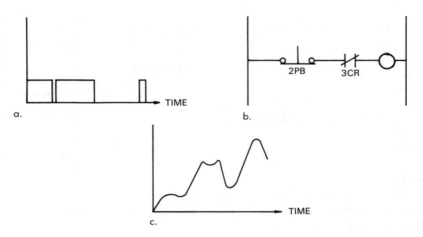

10-2.　The difference between analog and digital signals is that (a) digital (b) analog signals always change abruptly, as opposed to changing smoothly and gradually.

10-3.　Which system is more likely to be controlled by a programmable controller with analog capabilities?

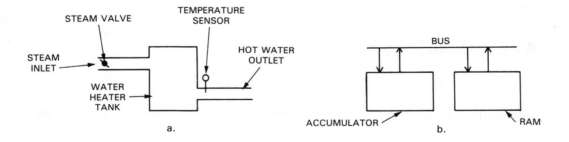

**Chapter 22
RELAY LOGIC AND PROGRAMMABLE
CONTROLLERS**

Name _____

Date _____ Score _____

Instructor _____

BOOLEAN ALGEBRA AND RELAY CIRCUITS

11-1. Write Boolean expressions for each of the following circuits. Do not simplify.

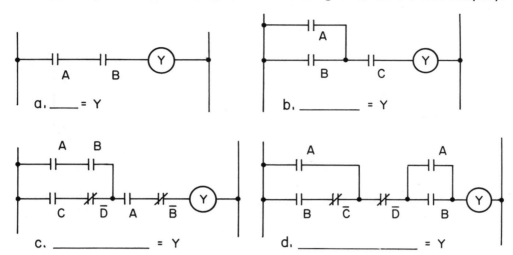

a. _____ = Y

b. _____ = Y

c. _____ = Y

d. _____ = Y

11-2. Draw relay-based circuits to represent the following expressions. Do not simplify.

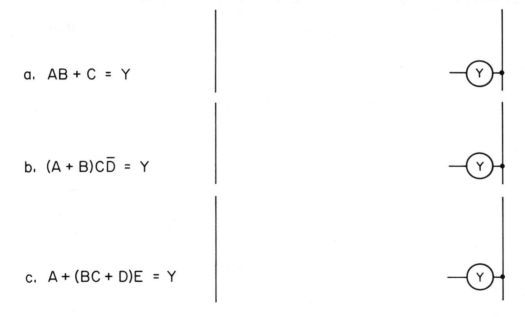

a. AB + C = Y

b. (A + B)C$\overline{D}$ = Y

c. A + (BC + D)E = Y

11-3. Which of the following circuits has memory capabilities? That is, which functions like a flip-flop?

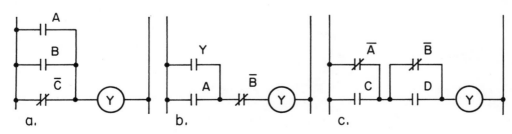

a.

b.

c.